Electrical Fires and Failures

Electrical Fires and Failures

A Prevention and Troubleshooting Guide

A. A. Hattangadi
Former General Manager
Chittaranjan Locomotive Works

McGraw-Hill
New York San Francisco Washington, D.C. Auckland Bogotá
Caracas Lisbon London Madrid Mexico City Milan
Montreal New Delhi San Juan Singapore
Sydney Tokyo Toronto

McGraw-Hill
A Division of The McGraw-Hill Companies

Copyright © 2000 by The McGraw-Hill Companies, Inc. All rights reserved. Printed in the United States of America. Except as permitted under the United States Copyright Act of 1976, no part of this publication may be reproduced or distributed in any form or by any means or stored in a data base or retrieval system, without the prior written permission of the publisher.

Previously published by Tata McGraw-Hill Publishing Company Limited, New Dehli, India, copyright © 1999.

1 2 3 4 5 6 7 8 9 0 DOC/DOC 9 0 4 3 2 1 0 9

ISBN 0-07-135651-7

The sponsoring editor for this book was Zoe Foundotos, the editing supervisor was Frank Kotowski, Jr., and the production supervisor was Pamela A. Pelton.

It was set in Garamond by Anvi Composers.

Printed and bound by R. R. Donnelley & Sons.

McGraw-Hill books are available at special quantity discounts to use as premiums and sales promotions, or for use in corporate training programs. For more information, please write to the Director of Special Sales, Professional Publishing, McGraw-Hill, Two Penn Plaza, New York, NY 10121-2298. Or contact your local bookstore.

 This book is printed on recycled, acid-free paper containing a minimum of 50% recycled, de-inked fiber.

Information contained in this work has been obtained by The McGraw-Hill Companies, Inc. ("McGraw-Hill") from sources believed to be reliable. However, neither McGraw-Hill nor its authors guarantee the accuracy or completeness of any information published herein and neither McGraw-Hill nor its authors shall be responsible for any errors, omissions, or damages arising out of use of this information. This work is published with the understanding that McGraw-Hill and its authors are supplying information but are not attempting to render engineering or other professional services. If such services are required, the assistance of an appropriate professional should be sought.

Contents

Foreword xiii
Preface xv

1. **Introduction** 1
   - 1.1 The Crying Need for a Book on Electrical Fires *2*
   - 1.2 Cure, Not Prevention, Gets Priority Today *3*
   - 1.3 The Approach and the Direction of This Book *5*
   - 1.4 Recurring Themes in This Book *8*
   - 1.5 Significance of Connector and Terminal Failures *10*
   - 1.6 Reasons for Failures of Electrical Equipment *11*
   - 1.7 Methods for Preventing Failures *13*
   - 1.8 Investigation of Electrical Fires and Failures *13*
   - 1.9 Management Has a Vital Role to Play *14*
   - 1.10 The Plan of This Book *16*
   - 1.11 Aspects Not Covered in This Book *16*
   - 1.12 Summary of the Main Points in This Chapter *17*

2. **Terminology** 19
   - 2.1 Introduction *19*
   - 2.2 Electrical Fire *21*
   - 2.3 Failure *21*
   - 2.4 Defect *22*
   - 2.5 Seed-Defects *23*
   - 2.6 Difference Between Defects and Seed-Defects *24*
   - 2.7 Modes and Mechanisms of Failure *25*
   - 2.8 Failure Rates *27*
   - 2.9 Inter-Relationships Between Terms *29*
   - 2.10 Metal Fatigue *29*
   - 2.11 Elasticity of Metals *30*

2.12 Metal Creep 30
2.13 Stress and Strain 31
2.14 Elastic Limit 33
2.15 Endurance Limit 33
2.16 Insulation Resistance and Dielectric Strength 34

3. Electrical Fires   35

3.1 Introduction 35
3.2 Scope of This Chapter 38
3.3 Causes of Electrical Fires 41
3.4 Fires Due to Failure of Insulation 45
3.5 Types and Causes of Insulation Failures 46
3.6 The Indian Electricity Act and Rules 48
3.7 Importance of Protective Systems 49
3.8 Protective Systems—Fuses 53
3.9 Protective Systems—Circuit Breakers 54
3.10 Prevention of Failures of Protective Systems 54
3.11 Fires Due to Failures of Pressure Contacts and Fractures of Conductors 55
3.12 Fire Prevention Measures 57
3.13 Conclusion 58
3.14 Do's And Don'ts for Preventing Electrical Fires 59

4. Transformer Failures   61

4.1 Introduction 61
4.2 Are Transformer Failures Due to Overloading? 64
4.3 Failure Modes of Power Transformers 66
4.4 Failures Due to Defects in Internal Connections and Terminals 68
4.5 Failures Due to Interturn Shorts in Windings 70
4.6 Failure of the Insulation Between Winding and the Tank 77
4.7 Other Failure Modes 78
4.8 Investigation of Transformer Failures 82
4.9 Procurement of New Transformers 82
4.10 Do's and Don'ts for Investigating Transformer Failures 83
4.11 Do's and Don'ts for Preventing Transformer Failures 83

5. Failures of Electrical Connectors and Terminals   85

5.1 Introduction 85

5.2	Typical Electrical Connectors 86
5.3	Apparent Triviality of Connectors and Their Defects 89
5.4	Effects of Failures of Electrical Connectors 92
5.5	Criticality of Failures of Electrical Connectors 93
5.6	Preventive Measures 93
5.7	True or Root Causes of Connector Failures 94
5.8	The Importance of Contact Force 95
5.9	Contact Force Developed By Screws or Nuts/Bolts 95
5.10	Contact Force Developed by Springs 96
5.11	Modes and Mechanisms of Electrical Connector Failures 96
5.12	Failures Due to Major Design Defects 98
5.13	Failures Due to Overheating 99
5.14	Do's and Don'ts for Preventing Failures of Electrical Connectors and Terminal Boards 102
5.15	Conclusion 102

6. Overheating/Burning of Crimped Sockets   104

6.1	Introduction 104
6.2	The Socket System of Terminating Cables 106
6.3	Tin Soldered Sockets or Lugs 106
6.4	Modes and Mechanisms of Socket Failures 111
6.5	Measures for Preventing Overheating/Burning of Sockets 115
6.6	Overheating Due to Terminal Board Failures 119
6.7	Wire Working Out of Socket 119
6.8	Do's and Don'ts for Preventing Overheating and Burning of Sockets 120
6.9	Conclusion 121

7. Failures of Plug/Socket Connectors   122

7.1	Introduction 122
7.2	Specification for Plugs/Sockets 125
7.3	Importance of the Contact Force 127
7.4	Three-pin Plugs/Sockets in Domestic and General Use 128
7.5	Industrial Multicore Couplers 131
7.6	Failure Modes 132
7.7	Overheating of Pin/Socket 133
7.8	Insulation Failures Between Adjacent Pins/Sockets 135
7.9	Open Circuits 135

- 7.10 Multicore Connectors in Electronic Devices *136*
- 7.11 Open Circuits Due to External Forces *136*
- 7.12 Precautions During Installation of Multicore Cables *137*
- 7.13 Summary *137*
- 7.14 Do's and Don'ts for Preventing Failures of Multicore Couplers *138*

8. Fractures of Crimped Sockets   139

- 8.1 Introduction *139*
- 8.2 Fractures of Crimped Sockets Due to Material Defects *140*
- 8.3 Fractures Due to Sharp Bends and Stress Raisers *142*
- 8.4 Fractures Due to Cracks Formed by Defective Crimp *144*
- 8.5 Fractures Due to Excessive Vibration *144*
- 8.6 Fractures of Sockets Due to Thermal Stresses *146*
- 8.7 Investigation of Fractures of Crimped Sockets *147*
- 8.8 Do's and Don'ts for Preventing Fractures of Sockets *148*
- 8.9 Conclusion *149*

9. Failures of Wire Strands   150

- 9.1 Introduction *149*
- 9.2 Causes Common to Socket Fractures and Wire Strand Fractures *151*
- 9.3 Heat Shrinkable Sleeves *152*
- 9.4 Strand Fractures Due to Sharp Edges on Socket Barrel *154*
- 9.5 Strand Fractures Inside Insulating Sleeves *156*
- 9.6 Do's and Don'ts for Preventing Strand Fractures *157*
- 9.7 Conclusion *157*

10. Failure of Insulation on Connector Cables   159

- 10.1 Introduction *159*
- 10.2 Insulation Failures Due to Mechanical Damage *161*
- 10.3 Insulation Failures Due to Thermal Damage *165*
- 10.4 Insulation Failures Due to Chemical Damage *167*
- 10.5 Insulation Failures Due to Electrical Damage *167*
- 10.6 Remedial Measures Against Damage to Insulation *168*

10.7	Do's and Don'ts for Preventing Cable Insulation Failures *168*	
10.8	Conclusion *169*	

11. Failures of Terminal Boards — 170

11.1	Introduction *170*
11.2	Failure Due to Inadequate Size of Terminals *172*
11.3	Failure Due to Inadequate Tightening of Threaded Fasteners *172*
11.4	Failures Due to Shrinkage of Terminal Boards *175*
11.5	Design for Reliability *178*
11.6	Importance of Contact Force *180*
11.7	Spring Washers *182*
11.8	Terminal Board Failures Due to Tracking *183*
11.9	Failures of Terminal Boards Due to Metal Creep *183*
11.10	Do's and Don'ts for Preventing Terminal Board Failures *184*
11.11	Conclusion *185*

12. Failures of Welded, Brazed and Soldered Joints — 186

12.1	Introduction *186*
12.2	Butt Welded Joints *188*
12.3	Tig Welded Joints *190*
12.4	Brazed Joints *191*
12.5	Soldered Joints *194*
12.6	Do's and Don'ts for Welding, Brazing and Soldering *195*
12.7	Conclusion *196*

13. Metal Fatigue — 198

13.1	Introduction *198*
13.2	Tensile Stress *199*
13.3	Stress Concentration Factor (SCF) *201*
13.4	Fractures Due to Metal Fatigue *204*
13.5	Alternating and Fluctuating Stresses *206*
13.6	The Nature of Metal Fatigue *207*
13.7	Causes of Fatigue Failures *208*
13.8	Appearance of Fatigue Fractures *209*
13.9	Do's and Don'ts for Preventing Fatigue Fractures *210*
13.10	Conclusion *210*

14. Metal Creep — 212

- 14.1 Introduction *212*
- 14.2 Effects of Metal Creep on Electrical Contacts *214*
- 14.3 Failures of Electrical Connections Due to Creep *217*
- 14.4 Prevention of Failures Due to Metal Creep *220*
- 14.5 Failures of Parallel Clamps *221*
- 14.6 Do's and Don'ts for Preventing Failures Due to Metal Creep *222*
- 14.7 Conclusion *223*

15. Electrical Contact Resistance — 224

- 15.1 Introduction *224*
- 15.2 Importance of Electrical Contact Resistance *225*
- 15.3 Factors Which Influence Electrical Contact Resistance *225*
- 15.4 Contact Theory in Brief *229*
- 15.5 Do's and Don'ts for Preventing Contact Failures *231*
- 15.6 Conclusion *232*

16. Shrinkage of Non-Metallic Materials — 233

- 16.1 Introduction *233*
- 16.2 Defective Design of Terminals *234*
- 16.3 Correct Design of Terminals *236*
- 16.4 Shrinkage of Insulation in Transformer Coils *238*
- 16.5 Slackening of Insulated Bolts *239*
- 16.6 Field Coil Failures Due to Shrinkage of Insulation *240*
- 16.7 Do's and Don'ts for Preventing Failures Due to Shrinkage of Insulating Materials *240*
- 16.8 Conclusion *241*

17. Thermal Degradation of Insulating Materials — 242

- 17.1 Introduction *242*
- 17.2 Characteristics of Thermal Degradation *243*
- 17.3 Insulation Failures in Electrical Connectors *245*
- 17.4 Do's and Don'ts for Preventing Thermal Degradation of Insulation *245*
- 17.5 Conclusion *246*

18. Electrical Tracking — 247

- 18.1 Introduction *247*
- 18.2 Tracking Failures on Terminal Boards *249*

18.3 Testing of Materials for Tracking Properties *249*
18.4 Prevention of Tracking Failures *250*
18.5 Do's and Don'ts for Preventing Failures Due to Tracking *251*
18.6 Conclusion *252*

19. Investigation of Failures and Fires   253

19.1 Introduction *253*
19.2 Fixing Priorities for Investigation *254*
19.3 Process of Investigation *256*
19.4 Failure Investigation: Step-by-step Approach *256*
19.5 Review of Failure Statistics *259*
19.6 Investigation of Electrical Fires *259*
19.7 Action Plan *261*
19.8 Conclusion *262*

Appendix 1   263
Fires Involving Electrical Equipment But not Classified as Electrical Fires *263*

Appendix 2   265
Technical Measures for Preventing Electrical Fires and Minimizing the Damages *265*

Selected Readings and Terms   268

Index   271

To

His Holiness Anandashram Swamiji
of Chitrapur Math, Shirali
N. Kanara, Karnataka

Foreword

This is an important book. And a very complete one. A. A. Hattagandi, former general manager for the Chittaranjan Locomotive Works of India's national electric railway system, has distilled his decades of experience with a broad range of electrical equipment and operating systems into a very thorough, very rigorous textbook on reliability engineering.

India's electric railways constitute a gigantic industrial and transportation enterprise spanning a subcontinent, complete with their own manufacturing plants, power generation facilities, transmission and distribution network, and maintenance operations. For this reason, they constitute a perfect laboratory for a broad and ongoing investigation into the causes of failures in electrical equipment and systems, and, more important, how such failures can be prevented in order to avoid downtime, prevent fires, and help avoid loss of life. Never limiting himself to transportation concerns, the author extends the technical and managerial knowledge gained through a career spent helping to keep the trains running on time (no small consideration in a nation with one-seventh of the world's people, most of them dependent on public transportation rather than private vehicles [and low rates of private vehicle ownership]) into a broad and perceptive analysis of the causes and prevention of electrical failures generally.

The author makes several important points in this book. The most original is a new concept that he calls a 'seed-defect,' meaning a basic flaw of design or workmanship that may cause equipment to fail. Such deficiencies, or seed-defects, which can occur in the design, manufacture, operation, or maintenance of electrical components and systems sometimes cause problems months or even years after a product or component is installed [put in service]. Only a few of these seed-defects 'germinate' and grow into flaws that many may cause failures or even fires. Since it isn't possible to predict which seed-defects will eventually cause performance or safety problems, the only safe course is to eliminate all of them. Fortunately, according to the author, this is both practical and cost-effective. Identification and elimination of seed-defects is therefore the key to achieving zero failure performance.

Another important point that A. A. Hattagandi explains is that 'material failure' and 'human failure' are fallacious causes used to explain away failures of equipment. As he amply demonstrates and explains, materials never fail without good reason. When a short circuit, ground fault, break in a conductor, or other failure mode occurs, it always happens in accordance with the laws of physics and measurable properties of materials. There is always a definite cause, a definite but avoidable deficiency, which causes the equipment failure [to fail].

Although this book is meant mainly for front-line supervisors and engineers, it would also make an excellent reliability engineering textbook. The author deals with complex disciplines as different as metallurgy, mechanical engineering, and electrical engineering. His text is clear, free of jargon, and amply illustrated. This book is truly international in scope and very successful. Not everything in it conforms strictly to U.S. [North American] practice. (For example, the author makes frequent mention of fires caused by temporary wiring systems in Indian marriage pavilions called *shamianas* and *pandals*.)

But the basic technical issues and principles that A. A. Hattagandi explains so clearly and thoroughly apply to the design and construction of electrical systems of all kinds, everywhere. Safety and reliability know no borders.

H. Brooke Stauffer
Director, Codes and Standards
National Electrical Contractors Association (NECA)
Bethesda, Maryland

Preface

During a career spanning four decades in the electrical department of the Indian Railways, I came across many cases of electrical fires and failures. I learned many lessons, but the most important one was that more than 50 percent of these fires and failures could have been prevented by improvements in the design, manufacture and maintenance of minor components such as terminals, connectors, threaded fasteners, cleats, washers, springs, fuses, etc. The remaining incidents were due to a variety of causes, but even these could have been prevented by improvements in one of the following stages:

- Specification and design
- Manufacture and installation
- Operation and maintenance

In short, with the current level of technology already available with us, there need not be any fires or failures in electrical installations or equipment. Zero failure performance is very much within our reach. Some professional electrical engineers who have to deal with the consequences of fires and failures may find this claim hard to believe. To them I can only say—read on!

All electrical equipment operates according to laws of nature such as Ohm's law, Newton's laws, Lenz's law, Faraday's laws, etc. All failures or fires also occur strictly according to these same laws and also some other laws of nature (Miner's law, Arrhenius' law, etc). There are no exceptions to this rule and there is never any stage in any failure process which is not in conformity with some immutable natural law. On this eternal truth rests the foundation of relability engineering. It is also the basis of my earlier assertion that zero failure performance is within our reach.

Electrical engineering courses in the 1940s did not give any guidance on reliability engineering and the position has not changed much in India even today. Over the years, I have been searching without success for a book on this subject which could be of help to the front-line electrical supervisors and engineers in the industry. There are, of course, books on reliability engineering but these are generally of no help to supervisors or engineers who have to deal with electrical equipments and hardware on the shop floor. The practical aspects of this subject are apparently considered to be too trivial to merit a book; but one continues to read in the newspapers about fires and transformer explosions involving loss of life and property. There are many more cases of failures of electrical equipment which largely go unnoticed. They not only affect services and production but also increase capital and maintenance costs.

What is more disquieting is that the explanations being given by those who should know better are often misleading and wrong. Overloading, worker indiscipline, heavy rain, thunderstorms, and short circuits are some of the reasons put forward by the authorities, for electrical fires and failures. Even cursory inspections of electrical installations in public buildings and in factories reveal many visible defects capable of causing fires and failures. Discussions with the technicians and junior supervisors show that no one has ever given them any systematic training on how to prevent electrical fires and failures. There is no comprehensive list of Do's and Don'ts.

After the occurrence of a fire or failure, the emphasis at the management levels is on rescue, fire fighting, repairs and the earliest possible restoration of services and resumption of production. If any formal enquiries are ordered, everybody including the local management goes on the defensive and the findings of the enquiry committee have no real effect. Similar incidents continue to take place.

No one seems to have either the time or the inclination to undertake in-depth investigations of the root causes of such incidents and there is no sustained drive to prevent their recurrence. There seems to be a tacit acceptance of the notion that fires and failures are bound to occur. Many seem to believe that sooner or later, materials will fail and people will make mistakes. Material failure or human failure are two overworked last words after the occurrence of fires or failures.

Preface

My experience in the electrical department of the Indian Railways has shown that wherever such incidents were investigated, the root causes determined and the required modifications implemented in the design or in the maintenance practice, there has been no recurrence of the types of failures dealt with. Many types of failures which used to occur frequently in the 1970s and the 1980s are now extinct. If further work is done on the same lines, zero failure performance is well within our reach.

The technology involved here is certainly not advanced and the preventive steps are inexpensive and well within our capacity. It is necessary only to learn to recognise or to identify the types of apparently trivial defects which grow, often insidiously and silently, into failures and fires. Our educational curricula, either for the engineering courses or for the industrial technician courses, do not give satisfactory coverage of the issues involved in this problem. There is clearly a need for a comprehensive book which explains in detail how and why electrical fires and failures are caused and what steps can be taken to prevent them. This book is aimed at fulfilling that need.

Any electrical installation of today consists of components and materials which have been manufactured, processed and installed by a large number of organisations located several hundreds if not thousands of kilometers from each other. The defect which is responsible for a fire may have been introduced at any one of these widely dispersed centers but the ultimate responsibility rests always with the last unit in the chain, i.e., the user industry. In any case, it is the user industry which has to suffer the consequences of fires and failures.

Special care has to be taken by the user in the drafting of specifications, selection of suppliers, inspection of materials and training of staff. Managements have a special role to play to ensure that all this is properly tied up. While this book is meant mainly for the front line supervisor and engineer, one chapter at the end deals briefly with the role of managements.

This book will prove useful for electrical supervisors, engineers and managers in all industries which generate, distribute or utilise electrical energy; and this means in effect industries of all types in mechanical, electrical, chemical and other disciplines because electrical equipments are used in all of them.

It is hoped that manufacturers of electrical equipment will also take note of the points made here. In their own long-term interest, they should aim at supplying items which do not fail and they should stop looking at supply of spares and service engineering as profit centers.

Electrical fires are rare for any particular industrial unit, but if and when they do occur, the consequences can be so serious that even insurance can only partially cover the total losses. This book aims to show that training the staff to take a few simple precautions can help the users to achieve total freedom from electrical fires and failures.

While electrical fires are rare, failures are not. Electrical failures are those incidents where some operations are interrupted and repairs generally call for replacement of only a few components. A disaster or even a fire may be averted either because some protective equipment has operated correctly or because the arc has got quenched by itself as very often happens in alternating current systems. Many such incidents do occur every day in almost every large industrial unit. They are not in the public eye except those which occur in the electrical distribution industry. Electrical failures are responsible for a significant part of the problem of electricity shortages, load shedding and electrical breakdowns.

The difference between electrical fires and failures is mainly with regard to the effects or consequences. These are determined largely by chance. The causes, the failure modes and the failure mechanisms (processes) are usually the same. The measures taken to prevent electrical fires will prevent at the same time a much larger number of electrical failures also. The cost of training the staff in this regard is very small in comparison with the cost of electrical fires and failures which will continue to occur in the absence of such training.

The information in this book will help in the drafting of course material for the training modules for artisan staff and the book itself should be a required reading for supervisors and instructors who have to train electrical artisans and for design engineers who deal with the design of small hardware and accessories.

Additional information and explanations have been included in the book to cover a number of useful comments and suggestions made by Shri J. N. Bhavani Prasad and Shri N. Venkatesan, who went through the manuscript and for which I am thankful to them. I am grateful to my wife and family for encouragement and support at critical times.

Criticisms and suggestions from readers will be most welcome.

A. A. Hattagandi
New Delhi

Chapter 1

Introduction

In this chapter

- 🔌 We will discuss some of the basic reasons for electrical fires and also take an overview of the measures that should be taken to eliminate them.

- 🔌 We will outline the approach to the problem, discuss the main themes which run through the book and describe its general plan.

- 🔌 The usual reasons for failures and fires of electrical origin will be introduced. Special reference will be made to the material degradation processes which are at the root of the problem.

- 🔌 The significance of electrical fires and failures originating from defects in connections and terminals, which are apparently trivial and consequently neglected, is discussed briefly. Chapters 5 to 12 discuss the same in detail.

- 🔌 As also in the book as a whole, the emphasis is on prevention of electrical fires and failures, with the aim of zero failure performance as an attainable and realistic target.

- 🔌 We briefly discuss the important role the managements play in regard to (a) the supply of high quality materials and tools to the production shops, (b) the training of staff to detect and to avoid defects in their work, and (c) to establish systems for the investigation of failures and monitoring of the implementation of measures to prevent recurrence of failures.

- 🔌 A broad perspective of the whole book is also given.

1.1 THE CRYING NEED FOR A BOOK ON ELECTRICAL FIRES

A few years ago, an accident involving a major fire in a marriage pavilion occurred. What started as a joyous occasion ended in a tragic accident involving the loss of many lives. The cause of the fire was described as an electrical short circuit.

At that time, I was conducting a course in reliability engineering. I took the opportunity to discuss this subject with the trainees who were junior engineers, each with about 10 years experience in maintenance of electrical equipment. Almost all the 15 trainees were of the view that an electrical short circuit could indeed be one of the likely causes of a fire.

This was shocking because the root cause of the fire had to be, in my view, quite different. A short circuit is an effect and not a cause. More precisely, a short circuit is a stage in one of several possible failure processes which end in a fire.

On further discussion, some of the trainees came out with some other possible root causes of the fire including poor quality of wiring, inadequate size of the wires, and overloading of circuits.

These causes were nearer the truth, but still far from the correct answer. All these are possible causes of an electrical short circuit or overheating of wires, but not of a fire.

An electrical short circuit has been said to be responsible for a number of other disastrous fires in many public places such as the Moore Market in Chennai, the New Market in Calcutta, a film set in Bangalore, a high-rise building in New Delhi, a marriage pavilion in Dabwali, Haryana, and the Sadar Bazaar in Dehli.

Subsequent discussions with a number of electricians, supervisors and engineers showed that there are many misconceptions among the general public, and also among many of those responsible for the design, manufacture and maintenance of electrical equipment and installations.

Many of the possible and highly probable defects which could lead to failures and fires are not known to many engineers, and no thought is being given to the measures that could possibly be taken to prevent the incidence of fires of electrical origin. The awareness about the potential dangers of electrical defects and failures is even less amongst the supervisors and the staff.

Introduction

The few books that are available on the subject neither sufficiently emphasise nor clearly explain why or how certain precautions should be taken. The need for a book on electrical fires and failures became apparent to me when a search of the bookshops revealed that a comprehensive book, which described at least the more probable causes of electrical fires and failures was not available. This was confirmed by many of the trainees in the reliability engineering course.

1.2 CURE, NOT PREVENTION, GETS PRIORITY TODAY

Reliability of equipment should be of prime concern to those in charge of the maintenance of plant and machinery in any industry. Failures of equipment create problems for the owners by way of losses in production or revenue, and the general public suffers due to interruptions or delays in essential services like electricity, water supply, and railways. When this happens, the supervisors and engineers in charge of maintenance come under pressure from all sides to restore the service as early as possible.

At present, the emphasis is only on quick repairs and restoration. Those in charge of maintenance often work round the clock under difficult conditions not only to carry out the scheduled or preventive maintenance on the entire installation, but also to carry out repairs and to restore the service when some of the plant or machinery develop snags and have to be shut down. Generally, there is no sustained effort thereafter to determine the root causes of the failure and to implement effective measures to prevent recurrence. There are several reasons for this apparent lack of interest in real preventive action. These are as follows:

- The maintenance supervisors and engineers are generally far too busy with repair works to be able to make the special effort which is necessary for investigating failures. By the time repairs are completed in one place, another failure of a different type in another place might call for urgent attention. The maintenance staff thus move around from one failure site to another and, even scheduled maintenance may get neglected. There is thus a double vicious cycle in operation here (see Fig. 1.1).
- Everyone has got so used to the idea of equipment failures occurring from time to time that there is a tacit acceptance of periodical failures as one of the 'facts' of modern life. Even top managements are generally not prepared to approve the

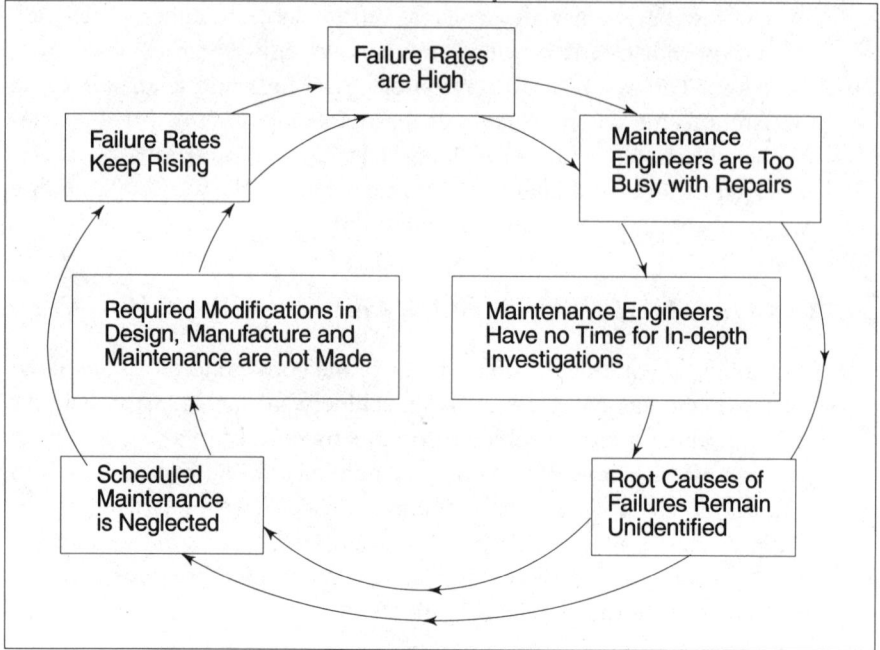

Fig. 1.1 Vicious cycle of failures and neglect

deployment of adequate resources to investigation and prevention of failures. They are satisfied as long as repairs are carried out quickly. And if, occasionally, they feel that something should be done to prevent such costly failures, the action taken is to advise the maintenance engineers to ensure that correct and timely maintenance is carried out. It is assumed that all failures are generally due to poor maintenance. Actually, less than 15 percent of the failures can be ascribed to deficiencies in maintenance.

- The majority of failures which take place are usually due to defects or deficiencies in design or manufacture of small details. The responsibility for design usually rests with the manufacturers of the plant, machinery, and their components; but this is rarely accepted by the manufacturers. Even investigations are not undertaken by them for short-sighted commercial reasons. Only if the problem is serious enough to attract the attention of the top management of the user industry, is there some investigation. All other so-called minor problems remain unattended. If the maintenance engineers complain about the failures to the

Introduction 5

manufacturers, their complaints are generally returned with suggestions about poor maintenance or maloperation. Often, the maintenance engineers stop making any complaints and manage as best as they can in the circumstances. The net result is that failures due to design defects continue to take place.

- Failure investigation always involves design considerations because the failure of any component is actually the outcome of the continuous tussle between the 'load'* and the 'strength'* of the component. Maintenance engineers are generally not trained to examine design aspects. As a result, they tend to shy away from any effort in that direction. It is easier and also more satisfying to concentrate on repairs because at the end of it, one can at least have the satisfaction of seeing the resumption of service.

This situation obviously cannot be allowed to continue. The public expectations of quality and reliability of services are increasing. Competitive pressures on the owners of industries are making them realise the need for improving the reliability of their plant and machinery. Equipment failures are far more expensive than the mere cost of repairs. Loss of revenue due to stoppage of production is usually very high. The day is not far when there will be zero failure expectation.

Given the current state of affairs, it is difficult for most people to appreciate that zero failure rates could ever be achieved. However, within the resources available to any average group of engineers and supervisors, it is indeed possible to attain zero failure performance in a reasonable period of one or two years. This book aims to initiate a drive towards this end through effective measures to prevent failures and fires.

1.3 THE APPROACH AND THE DIRECTION OF THIS BOOK

The approach in this book to the problem of equipment failures is through a study of what might be called design principles. However, reading and understanding its contents should not call for any prior knowledge of mathematics, statistics or engineering design. The book is meant mainly for supervisors, foremen and junior engineers who have practical experience but no engineering qualifications. Most engineering principles are actually based on common sense; and it is

---

* The terms load and strength are used in a general sense. They could be either electrical or mechanical parameters such as breakdown voltage, tensile stress, temperature, etc.

necessary to know only these basic principles to understand how or why failures take place and to appreciate the need for certain precautions that have to be taken.

Advanced theories and mathematics are required only at the stage of original design. Although this book is not intended to explain how to design electrical equipment, a few electrical engineering design principles are discussed, keeping in view the capabilities and the needs of those people in the field. At the same time, design engineers should find many useful practical aspects in this book to help them to design intrinsically reliable equipment. The approach to failure analysis through design principles is as useful to maintenance engineers as to design engineers.

This book should also prove to be useful for young graduate engineers who find themselves in charge of the maintenance of electrical and mechanical equipment. Our university courses usually teach how things work but not how or why they fail. Even those who have taken optical courses on reliability engineering find that the practical world of fractured components and burnt equipment is far removed from the world of statistics, hazard functions and Weibull charts. Although this book is about reliability engineering, its approach is somewhat different from the academic or mathematical one that is followed in books on the subject.

The usual emphasis in books on reliability engineering, is on the statistical, mathematical and academic aspects of reliability. These aspects are certainly of great importance for those who want to understand the fundamental theories, to explain certain observed relationships between failure rates and contributory factors and to interpret failure statistics correctly. Such knowledge is essential in the design of equipment, design of inspection and quality control procedures in mass production and in investigation of failure data. However, these methods can be applied only when dealing with large volumes of production or large numbers of failures of equipment of identical design. In the chapters which follow, the emphasis will be on failure mechanisms and material degradation. This approach will enable the reader to determine the root causes of failures. It will help him to deal with problems where failure statistics are not available, as is often the case, and reliability theories cannot be applied.

Although various industries such as the railways, manufacturing industries, and thermal power stations are quite different from each other with regard to the technology involved in their installation and

Introduction

operation, there are a number of common elements when we consider the individual equipment and components of the plant and machinery. There is an even greater degree of commonality between the modes, mechanisms and causes of equipment failures generally experienced in these diverse industries. This is not surprising because the materials used in the construction of the plant and machinery are, by and large, the same in all these industries, and generally include various grades of iron and steel, copper, aluminium, and their alloys, plastics, elastomers, ceramics, and laminates. Our approach to the problem of failures through the study of material behaviour will, therefore, prove useful in all the industries.

Often, failures continue to take place because the preventive steps taken are totally ineffective. Sometimes, failure rates actually increase after taking what are thought to be preventive steps. These unfortunate results follow the action taken only when the cause of failures has not been identified correctly.

When the average supervisor or engineer, who is in charge of the maintenance and repair of electrical or mechanical equipment is faced with their failures, the available books on reliability engineering offer him no help to determine the causes of the failures or to evolve the remedial measures for preventing similar failure is from recurring.

This book does not cover the repair of failed equipment. Our supervisors and engineers are generally quite good at carrying out repairs. They often do this with great speed and at minimum cost. Innovative methods are evolved and round-the-clock work is done to restore normal production or service. Though there may be some scope for a study on advanced repair techniques, our emphasis here will be on prevention and not on cure.

Usually, in any industrial unit, the failure modes encountered from day-to-day are not the same; and even the equipment or components which fail are not the same. The usual pattern is: one or two cases every year of each type of failure. The number of types of failures, however, large enough to make the total failure rate of the system so high as to create anxiety on account of their adverse effect on production or service.

Even where the number of failures of the same type is substantial—say 10 or more per year, the statistical and mathematical methods taught by the books on reliability engineering are generally of no avail unless full and accurate data of different variables are recorded carefully. Such data are usually not available. In these circumstances, the statistical and mathematical methods of reliability engineering cannot be applied but the approach followed in this book will surely give results.

To determine the failure modes, the mechanisms of failures, the root causes of failures and the remedial measures under these conditions, it is necessary to adopt an altogether different approach. This book is an attempt to show a new direction. It aims to show practical steps that can be taken by the person in charge of the equipment to determine the true cause of the failure and the preventive measures to be taken in regard to electrical connectors, terminal boards and other vulnerable components.

1.4 RECURRING THEMES IN THIS BOOK

In this book, fires and failures caused by various types of defects in electrical equipment will be discussed and their root causes as also the remedial measures will be dealt with. The recurring themes which run through this book may, however, be stated thus:

(a) An electrical short circuit is never the cause of an electrical fire. It is only a stage in the process which starts with an insulation failure and ends in a fire. The real or root cause is usually a defect or deficiency introduced at some stage in the protective equipment either in design, manufacture or maintenance.

(b) It is possible for an electrical fire to start even without passing through the stage of a short circuit. In such cases, the defect is usually a bad contact or a fractured conductor. There is neither an insulation failure nor a short circuit, at least in the initial stage.

(c) Another group of possible causes of fires involving electrical equipment has nothing to do with any defect in the design or manufacture or maintenance of electrical equipment. This group covers fires due to the misuse of electrical appliances. Such cases are not discussed in this book because the possibilities are limitless and the remedies are obvious. However, a few of these cases are enumerated in Appendix 1.

(d) Electrical failures are the precursors of fires. Most electrical failures have the potential of developing into fires. The causes as also the preventive measures are the same for both.

(e) Electrical failures as also electrical fires of either type (a) or type (b) above can be totally eliminated. The measures to be taken are very simple and practicable. The costs involved in preventive action are negligible as compared with the savings that would be effected by the prevention of such failures and fires.

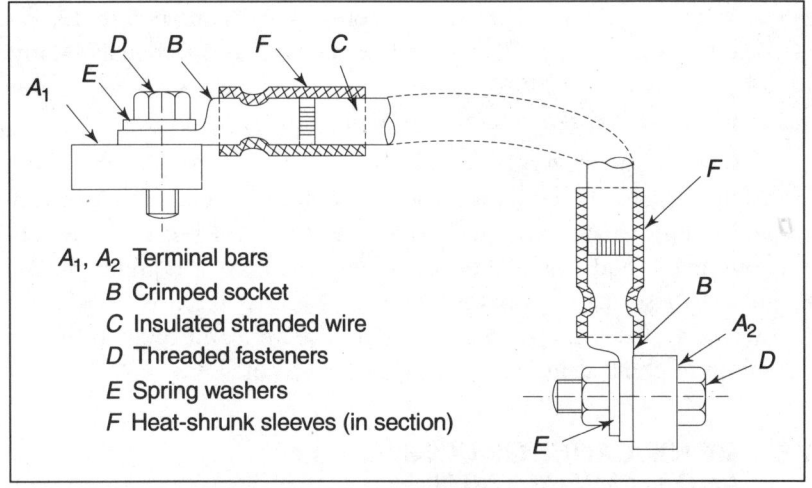

Fig. 1.2(a) Typical electrical connector

Fig. 1.2(b) Typical control panel showing the number of connectors

A significant part of this book, comprising Chapters 5 to 12, is devoted to the failures of electrical connectors and terminals. Many of the defects as described therein would at first appear to be trivial, which is what makes them more dangerous.

Whereas many types of defects in electrical equipment cause the equipment to merely stop working, defects in electrical connectors and terminals can and often do lead to fires which destroy the entire equipment and sometimes even other adjacent equipment. What makes these defects doubly dangerous is that there is no protective system available as yet, which could raise an alarm or shut off power supply automatically when such defects occur.

1.5 SIGNIFICANCE OF CONNECTOR AND TERMINAL FAILURES

Eight chapters in this book are devoted to the failures of electrical connectors and terminals because a very large number of failures and fires of electrical origin are due to apparently trivial defects in these components.

A typical electrical connector, as shown in Fig. 1.2(a), is a very simple component. What makes it important is that although it is required in very large numbers as seen in Fig. 1.2(b), each connector is prepared and installed manually, often by untrained staff. Figure 1.2(b) shows a panel with over 50 electrical connectors and over 200 pressure contacts. Each one of these is a potential source of failure if there are any defects or deficiencies in the hardware or in the workmanship.

Perhaps it is the very simplicity of electrical connections which is responsible for the inadequate attention being paid to their design, manufacture and installation. As a result, failures in service of electrical connectors are not as rare as may be expected. What is worse, many equipment failures which are often ascribed to defects in the more complex insides of the equipment are actually due to defects in electrical connections. Components which account for barely 5 percent of the initial cost are often responsible for 50 percent of the failures in service.

There are many books available in the market on the design of electrical equipment and machines; but these deal with only the main components such as the windings and their insulation, and the magnetic core. There is little or no mention of electrical connections—either internal or external, inside, or between different items of electrical equipment.

It is not surprising, therefore, that failures of electrical connectors are not rare. The neglect at the design and manufacturing stages is matched by the inadequate investigations which follow the failure of an electrical equipment. Since, usually, all that can be seen at the site of the failure is charred insulation, ashes and chunks of molten and congealed metal, it is natural that the engineers and supervisors in charge should consider it more useful to concentrate on repairs and rehabilitation than to undertake what seems to be a hopeless investigation.

When such a failure occurs, the technician or worker who made the connection may often get blamed for the failure, but even that conclusion is not pursued because there is no way to prove his negligence. Besides, there could be many other possible reasons which could lead to a similar outcome.

Bad fitting or assembly may well be the actual cause of failure, but it is often not considered significant to describe the mistakes made by the artisan. The artisan is neither trained nor guided correctly with regard to the do's and don'ts while making electrical connections. The subject is considered too trivial to need any mention.

Actually, more often than not, the true cause of failure is a defect in the design or manufacture of the components or tools used in making the electrical connection, but it is rarely so pin-pointed. In such cases, an artisan may be blamed unjustly. Either way, the true cause of failure is not established, the correct preventive action is not taken and similar failures continue to take place.

1.6 REASONS FOR FAILURES OF ELECTRICAL EQUIPMENT

The reasons for failures of electrical equipment can be classified into two major groups:
- The first group includes failures due to intrinsic weaknesses introduced in the equipment during either design, manufacture or installation. In such cases, the failure rates are high in the early life or the warranty period of the equipment. Generally, failures in this group get fully investigated and suitable corrective or preventive steps are taken by the manufacturers.

- The second group covers failures which are the result of certain degradation processes that are responsible for changes in the properties of materials as a result of the passage of time, effects of internal stresses and action of external environmental factors. Deficiencies, of a type different from those mentioned above, in design, manufacture, operation and maintenance can initiate and accelerate the degradation processes and cause failures in service at any time during the life time of the equipment.

The majority of electrical failures and fires are due to reasons in the second group mentioned above. Before one can begin to diagnose their root causes, it is necessary to understand fully the common degradation processes and failure mechanisms. Hence, a few chapters have been devoted to the study of these physical and chemical phenomena involved in the failures of all mechanical and electrical hardware or equipment.

Six physical phenomena involved in the processes and mechanisms of failures of electrical components are discussed in this book. They cover more than 90 percent of the failures which occur in practice and include:
- Metal fatigue
- Metal creep
- Electrical contact resistance
- Shrinkage of non-metals
- Thermal degradation of electrical insulation
- Electrical tracking

It will be observed that all these physical processes are actually involved in degradation processes which progress with usage of the equipment and the passage of time. They help us to explain how or why a component which was working well enough for months or even years, should suddenly fail while in service. When we know the mechanisms of failure, it is easy to determine the correct preventive action. Each of these degradation processes is covered by a vast amount of literature. In this book, only a very brief summary of the main features of the process are dealt with to the extent necessary to appreciate the suggested measures to prevent failures.

This book aims to give only a bare introduction to each of the six degradation processes mentioned above, but in each case, practical examples will be described in sufficient detail to enable the reader to grasp the fundamentals and to apply them effectively to investigate and to solve his own problems.

There are many more degradation processes which lead to failures of electrical equipment; but some of them are generally visible and better known. Two such processes are corrosion and wear. However, it can be stated that more than 90 percent of the failures which take place are due to one or more of the less obvious but more important, degradation processes discussed in detail here.

1.7 METHODS FOR PREVENTING FAILURES

It would be evident from the discussion so far, that the best way to prevent failures in electrical equipment is to identify the degradation processes responsible for the failures and then to take measures to prevent the onset, growth and acceleration of the degradation processes. These will be discussed in the following chapters.

It is not always practicable to control degradation of materials, particularly when one is saddled with large numbers of expensive equipment which has built-in, design or manufacturing deficiencies. In such cases, one has to resort to other methods of preventing failures.

When an equipment actually fails, some component is found to have failed by cracking, fracture, burning, or melting, but long before this happens, certain initial signs do appear sometimes. It is not always easy to detect these initial signs of impending failure, but it is possible to do so with the help of special condition monitoring equipment and methods. This is one of the ways of preventing failures during service, in equipment which is already vulnerable due to the presence of various defects. Some of these methods will be discussed in this book.

1.8 INVESTIGATION OF ELECTRICAL FIRES AND FAILURES

It is unfortunate but true that most manufacturers of electrical equipment try to avoid any involvement with failures of their products in service. Complaints from the users are usually disposed of without making any effort to determine the root causes of the failures. While it is hoped that attitudes will improve with the advent of global competition and user enlightenment, it will be decades before any effects are noticeable. In the meanwhile, the investigation of electrical fires and failures is a responsibility which should be undertaken by the maintenance supervisors and engineers.

Moreover, maintenance people are in the best position to carry out investigations because they would be the first to deal with the failed equipment. They are also the direct beneficiaries of successful preventive action determined by the investigation. Failures tend to occur repeatedly when they are due to design or manufacturing deficiencies. This can lead to adverse effects on production or service. Normalcy, i.e., zero failure performance, can be restored only when the deficiencies are identified and suitable corrective action is taken.

The investigation of failures and fires can be difficult and frustrating when the root causes are not identified. On the other hand, it can be interesting and even exhilarating if it is successful and the incidence of failures of a particular type can be swiftly arrested.

The best method of arresting failures is by preventing the onset of the degradation processes through the implementation of judicious changes in the design of the equipment. To be able to do so, it is first necessary to decide which of the various degradation processes are actually at work and then to examine how they can be either prevented from starting or slowed down sufficiently to prevent failures from taking place in service. Once these questions are answered, the remedial measures usually become obvious at once. All this is an essential part of any investigation into failures. Chapter 19 will deal with this subject in greater detail.

1.9 MANAGEMENT HAS A VITAL ROLE TO PLAY

The implementation of the measures for prevention of failures and fires in electrical equipment involves several management issues. Only the active participation of the management can result in the achievement of enduring results. Action has to be taken in an appropriate and coordinated manner in all the concerned departments or sections of an organization. Although this book is not about management principles in general, the specific actions that need to be taken by managements for realising the objective of zero failure performance have been discussed in the last chapter of this book.

There are several different but complementary reasons for inviting the special attention of managements and administrations to the problem of electrical fires and failures. These include the following:
- A disproportionately large number of failures of electrical equipment are actually due to apparently trivial defects in the design, manufacture and maintenance of electrical connections.

Introduction

- Electrical fires can be totally eliminated by taking a few simple precautions. This effort will also greatly reduce the number of electrical failures because the contributory causes of both fires and failures are usually the same.
- There is no comprehensive book available on the design, manufacture, maintenance and failure investigation of electrical connectors and terminals.
- Inadequate attention is being paid, even by reputed manufacturers of electrical equipment, to certain vital details relating to the design of electrical connectors and terminals. There have been many instances of failures and fires in equipment manufactured by reputed manufacturers abroad, due to defects in the design of connectors and terminal boards.
- The prevention of failures of electrical connections has the lowest cost/benefit ratio. The cost of achieving zero failure performance of electrical connections is negligible. On the other hand, the cost of failures of electrical connections is generally very high. Sometimes, when the result is a fire, the cost of property and lives lost can be horrendous.

While the causes of failures and fires are defects in the electrical equipment, there are many managerial problems, particularly in large organizations, which need to be solved in order to prevent the occurrence or recurrence of defects. Investigations into failures and fires must therefore determine the technical causes of the failures or fires and then go on to determine the organizational lacunae which led to the incidence of these technical causes.

Supply of high quality tools and materials, and the deployment of trained staff to the production shops are very important functions of the management. Material purchase policies have to be carefully formulated. Adequate resources must be allotted to the training of artisan staff in general, and to the development of training course material, training aids and training of trainers, in particular.

A few other important aspects of the role of the management may be summarized here as follows:

- Management must send a clear message or signal down the line that it expects zero failure performance.
- Management must deploy dequate resources to the investigation of failures and to the implementation of preventive measures.

- Management must monitor continuously the failure reporting, investigating and corrective action systems.
- Management must recognise and reward successful efforts resulting in prevention of failures.

1.10 THE PLAN OF THIS BOOK

Chapter 2 begins with the definitions of certain terms which are used repeatedly in the rest of the book. The terms are actually well-known, but it is necessary to define them carefully at the outset and to emphasise the differences between words which are apparently synonymous.

Chapter 3 is about electrical fires in general. The three main groups of defects which lead to electrical fires are discussed in detail. Various types of defects are enumerated, along with references to the other chapters where they are discussed in detail.

Chapter 4 is about transformer failures. A separate chapter is devoted to this equipment because a significant number of transformer failures are due to reasons other than defects in electrical connectors and terminals.

Chapters 5 to 12 are about various types of defects in electrical connectors and terminals which are responsible for many failures and fires in the majority of electrical hardware and equipment in factories, industrial plants and transportation systems of all types.

Chapters 13 to 18 deal with a number of degradation processes. This subject is also discussed in general in Section 1.6.

Chapter 19 deals with the investigation of failures.

1.11 ASPECTS NOT COVERED IN THIS BOOK

There are, in addition to failures and fires of electrical origin, failures of electrical equipment which do not involve electrical phenomena. Examples of such failures would be bearing failures, fractures of springs, gears, shafts, seals, and structural components. Failures of this type are well covered in the available published literature. This book covers the electrical failure modes which have not received adequate attention so far.

While the majority of failures in real-life situations are the result of some degradation process, there are a few cases where failure may occur due to intrinsic weaknesses in the equipment or overloading of the equipment. Such failures are relatively rare and they usually occur very early in the life of an equipment, soon after a new design is commissioned for the first time. In such cases, changes in design may become necessary, and where this is not practicable, operating conditions may have to be modified. Examples of such failures will be discussed briefly.

In addition, as already mentioned, the following aspects are not discussed for reasons indicated in the relevant sections:
- Statistic, mathematical and academic aspects of failure analysis
- Fires and failures caused by misuse of electrical appliances
- Methods and tchniques of repair of failed equipment

1.12 SUMMARY OF THE MAIN POINTS IN THIS CHAPTER

Electrical fires are the results of:
- defects or deficiencies in the protective system or
- defective pressure contacts or
- fractured conductors.

Electrical short circuits are not the causes of fires. They are stages in the failure processes which sometimes, but not always, end in fires.

Insulation failures and short circuits can be minimised, but they cannot be totally prevented; but fires can be prevented.

Apparently trivial defects in connectors and terminals are responsible for most cases of failures and fires; and they can be prevented by taking a few simple precautions during design, manufacture and installation.

Zero failure performance of electrical equipment can be attained through:
- identification and elimination of intrinsic deficiencies in design and control of degradation processes
- training of staff to identify and to avoid a few types of seeding defects and deficiencies
- in-depth investigation of failures and fires followed by an action plan to implement the required changes in design, manufacture and maintenance.

Management has an important role to play by:
- Organizing the training of staff in correct maintenance practices and elimination of seed-defects.
- Arranging proper tools, condition monitoring equipment and high quality materials.
- Deploying adequate resources for investigation of failures and monitoring of corrective action.
- Aiming at zero failure performance.

Chapter 2

Terminology

> In this chapter, we will discuss:
>
> ⚡ The meanings of certain special words or technical terms used in this book.
>
> ⚡ The terms discussed in the chapter include:
>
> ❑ Fire, Failure, Defect, Seed-defect.
>
> ❑ Failure Mode, Failure Mechanism, Failure Rate.
>
> ❑ Stress, Strain, Elastic Limit.
>
> ❑ Metal Fatigue, Metal Creep.
>
> ❑ Insulation Resistance, Dielectric Strength.
>
> ⚡ The relationships among the terms seed-defect, defect, failure and fire (shown in Fig. 2.1).

⚡ 2.1 INTRODUCTION

In this book, the following issues faced by organizations and the staff responsible for the operation and maintenance of electrical equipment and who consequently are answerable for their failures, are discussed

- The root causes of failures and fires
- The factors which contribute to their occurrence
- The physical processes involved in their development
- The measures which can be taken to prevent them

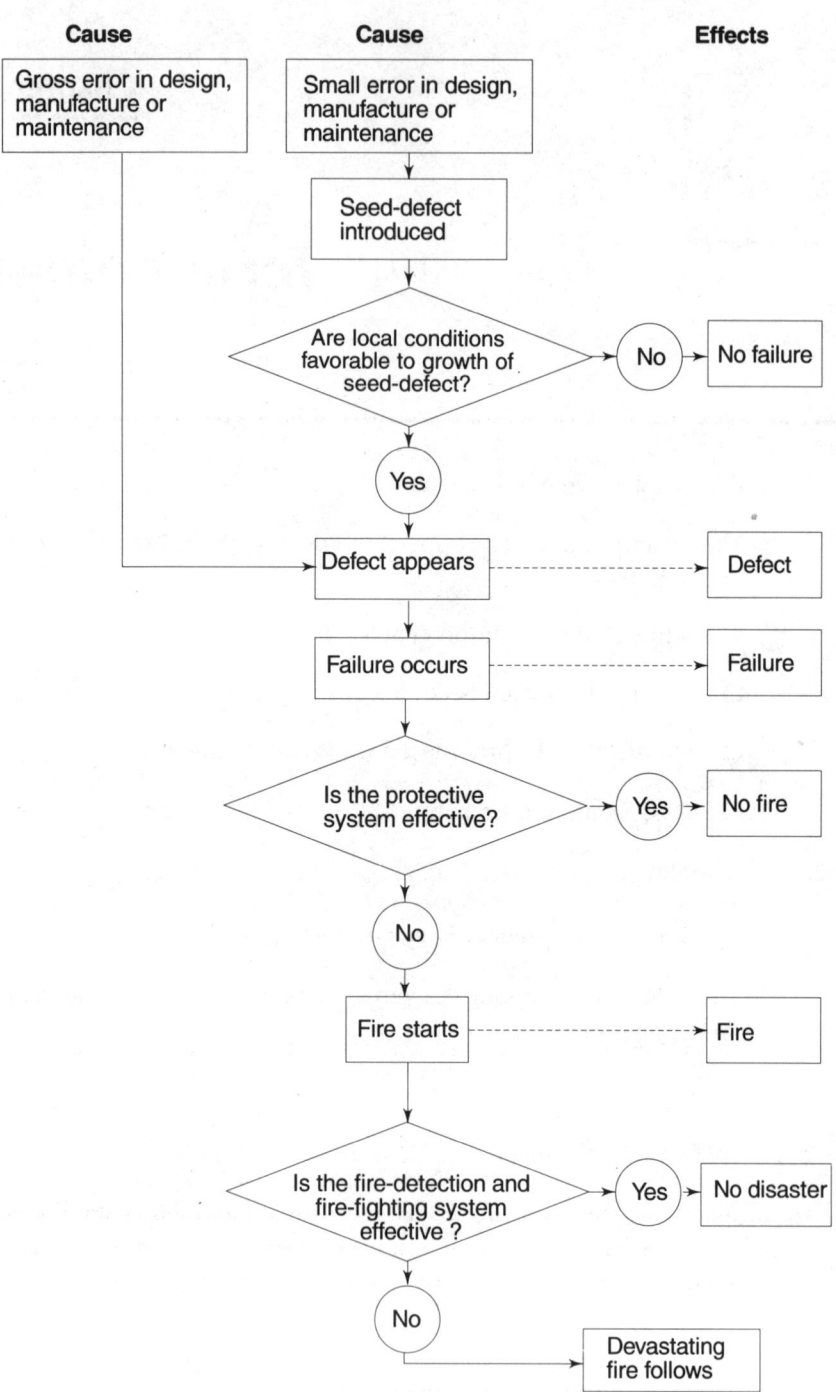

Fig. 2.1 Flow chart showing how seed-defects develop into fires and failures

- Before taking up a detailed discussion on the above lines, it is desirable to define and explain various terms used in this book. Many of these terms are commonly used words. Yet, it is necessary to define them clearly, because their meanings may not necessarily be the same as when they are used in engineering literature.

2.2 ELECTRICAL FIRE

The term electrical fire as used in this book will cover cases in which electrical faults lead to overheating and/or arcing, followed by ignition and combustion of insulating material in the equipment itself and sometimes, also of combustible material in the building which houses the equipment.

Some examples of electrical fires are:
- Fire on a switchboard panel caused by a defective terminal connector.
- Fire in a circuit breaker caused by a defect in its mechanism.
- Fire in a pavilon caused by a defect in the electrical protection system.

2.3 FAILURE

The term failure will cover the cases in which the electrical equipment fails to perform its normal function as a result of some electrical fault. In these cases also, there may be momentary arcing and local overheating, but there would have been no ignition and combustion. The power supply would have been automatically and almost instantaneously disconnected by the protective system. In a failure, there is interruption or stoppage of service, but the damage to equipment is minimal. In some cases, the damage may even be invisible.

If the protective system is defective or non-existent, any electrical failure is almost certain to develop into a fire. It has to be emphasised that an electrical failure is invariably the precursor of an electrical fire. If the failure is not detected quickly by a suitable protective system and power supply to the affected part is not disconnected at once, there is a fire. It is necessary to understand the distinction between these two terms clearly.

It is possible to minimize the number of failures, but it is extremely difficult to eliminate them completely. However, it is certainly possible to completely eliminate fires of electrical origin, as the number of ways in which they can start is small and many protective systems are easily available.

2.4 DEFECT

The term 'defect' refers to the proximate or immediately preceding cause of a failure. It may be in the form of a visible or measurable physical condition. The defect may be detected during normal maintenance before it develops into a failure during service. In fact, the objective of all maintenance or inspection schedules or programmes is to detect and to eliminate defects before they develop into failures. However, it is not always possible to take such action in time, and failures during service are then the inevitable result.

Some common examples of defects are given below:
- Damaged insulation on the wires due to abrasion or moisture.
- The use of a wire of inadequate size.
- A stray metallic wire or object coming in contact with bare unprotected terminals.
- Wire insulation which has cracked or deteriorated in some other way due to ageing.
- A vibrating conductor which has developed hairline cracks due to metal fatigue.
- A threaded fastener which has become loose as a result of vibration.
- A relay or miniature circuit breaker which has become jammed due to accumulation of dust.

The defects mentioned above may or may not affect normal operation of the equipment immediately; but sooner or later, they are sure to cause failures of the equipment in service. The time interval before this occurs will depend on a number of factors such as intensity of utilisation, severity of the defect, and environmental conditions.

There are also many other types of possible defects which can lead to failures and perhaps to fires. In the following chapters, we will discuss in detail, the various types of defects which might be present in an electrical installation, the measures which can be taken to prevent the occurrence of such defects, and also the modes and the mechanisms of failures which follow the occurrence of defects. These new terms are defined in the following sections.

2.5 SEED-DEFECTS

Most technicians and engineers recognize a defect which could lead to a failure, either immediately or within a short time. For instance, if a cable with visibly damaged insulation is allowed to remain in service, it could lead to a short circuit and perhaps a fire. Gross defects of this type do occur from time to time, but failures due to them are rare because the maintenance staff usually recognise and eliminate them quickly.

The majority of failures and fires which occur in practice are those caused by defects of a different kind, which we shall call seed-defects, i.e., defects which develop slowly or lie dormant harmlessly for years without causing any problem. If, for instance, the cable insulation is bearing hard on a structural member, it may be neglected. This is an example of a seed-defect which may or may not develop into a defect and later into a failure or fire.

One of the reasons why failures often take place on account of apparently trivial and avoidable seed-defects is the element of low probability which comes up in their activity. The presence of these seed-defects or embryonic defects does not necessarily cause immediate or even eventual failures. Out of a 100 components with a particular type of seed-defect, 90 pieces may not fail even after 20 years of service, but two components may fail within five years, and two more in the next five years, and so on. While the probability of any particular seed-defect growing into a full-fledged defect/failure/fire may be very low, the number of such seed-defects is generally so high that the overall probability of a fire or a failure is not negligible. It is not difficult to detect and even eliminate such seed-defects, but this is possible only when they are recognised as the root cause of the problem.

While it is not possible to predict whether or when a seed-defect in a particular component would lead to a failure in a particular equipment, it is possible to predict that some failures would certainly occur in a large population of similar components with that particular seed-defect. It is not unusual to see many installations working without failure, despite the visible presence of certain seed-defects. This fact is also responsible for the indifference of many workers and supervisors to the presence of such seed-defects. Elimination of these seed-defects is neither difficult nor expensive, and once it is done, it is possible to achieve zero failure performance.

The seed-defect is often, though not always, the starting point of a defect. This concept will be discussed later in greater detail, but suffice it to say now that seed-defects may not look like defects, or they may even be invisible; they generally lie dormant for months or even years. Some of them may, when the conditions are favourable, 'germinate' and develop into defects and then into failures or perhaps further into fires. Seed-defects grow insidiously and silently, and then strike suddenly when least expected.

Defects in protective devices should also be considered as seed-defects in the system. Protective devices operate only when there is a fault. A defect in a protective device will therefore remain undetected. There would be no problem and operations will continue normally as long as there is no fault in the protected circuit; but if and when a fault does occur, a fire and perhaps a disaster may follow.

2.6 DIFFERENCE BETWEEN DEFECTS AND SEED-DEFECTS

As many readers may have come across the concept of seed-defects for the first time, it is desirable to clarify their characteristics by comparing them with defects.

Defects are usually visible or measurable. A bolted connection which has not been tightened fully is defective. This defect can be detected by trying to tighten the screw or nut with a torque appropriate to its size. If the screw or nut turns through even a small angle, it means that the original fitting was defective.

It is, however, possible for bolted connections to have seed-defects even when the original tightening is correct. For instance, some part of the assembly may be subject to metal creep or some insulating material under compression may be subject to a gradual but permanent contraction. This could be a very slow and minute process which takes months or even years before its cumulative effect manifests itself, first as a defect and then as a failure or fire. The manner in which a seed-defect grows into a defect is explained in Chapter 11.

Only a few of the seed-defects present in an installation may grow into defects and then more quickly into failures or fires. It is possible that the majority of seed-defects in an equipment may never turn into defects. It is this characteristic which makes seed-defects doubly dangerous. Maintenance staff are likely to view such seed-defects with complacence. It is therefore necessary to train the maintenance staff to recognise the seed-defects and to eliminate them.

Terminology

The total number of types of seed-defects which can occur in electrical connectors and terminal boards is less than 20. When attention is drawn to these seed-defects and the manner in which some of them may grow into defects, failures and fires is explained to the staff, they will find no difficulty in eliminating the seed-defects. Zero failure performance can thus be achieved at practically no additional cost. Since seed-defects can be introduced in design, manufacture and maintenance stages, the staff in all these sections must be trained in this regard.

Every defect, sooner or later, will develop into a failure or a fire; but only a few seed-defects will do so. Most of the seed-defects will remain harmless during the life-time of the equipment. Since it is not possible to predict which seed-defects will create problems and which will remain harmless, it is necessary to identify and eliminate all of them.

The difference between a defect and a seed-defect may be further clarified by the following comparison relating to roller bearings:

Defects	*Seed-defects*
• Lack of lubricant	• Contaminated lubricant
• Severe pitting and spalling of races	• Electric current through bearing
• Bearing clearance nearly zero after installation. (Bearing failure is certain, sooner or later)	• Bearing clearance 80% of specified clearance. (Bearing may or may not fail, depending upon many unknown factors)

2.7 MODES AND MECHANISMS OF FAILURE

The terms 'modes of failures' and 'mechanisms of failures' will repeatedly appear in this book. It is therefore necessary to explain their meanings as used here, because the distinction between them is important.

The term 'mode of failure' covers the visible or measurable changes in the properties or appearance of the component which fails first during the process which precedes an equipment failure. It does not cover observations on the adjacent components damaged as a consequence of the first failure. However, in the initial stages of investigation, when it is not clear as to which component failed first, it is advisable to record observations on all the components in the damaged zone. Some examples of failure modes are:

Component	Failure mode
wire insulation	punctured
wire insulation	charred or burnt
copper conductor	melted
copper conductor	fractured
resistor	open circuit
capacitor	short circuit
shaft	fractured
ceramic insulator	flashed over

The term 'mechanism' normally brings to our mind a mechanical device comprising levers and gears, or the manner in which any device works. However, the meaning is quite different when we talk of a 'mechanism of failure'. In the context as used here, the term 'mechanism' refers to an internal physical or chemical process which ends in a failure of the material, component or equipment. For instance, a certain type of failure may involve physical processes such as thermal expansion, plastic deformation, elastic stress or fatigue in materials. In another type of failure, phenomena such as oxidation or heating effect of electric current may have each a specific role to play.

The important point here is that in every case of failure, there is a certain sequence of invisible processes, which may be either physical or chemical, the end point of which is the failure as we see it. It is only when we determine and understand the sequence of processes, or the true mechanism of failure, that we are able to take effective measures to prevent the failures from taking place. Obviously, it is necessary to eliminate the factors which start, support or accelerate each of the processes involved in the failure mechanism, if we are to succeed in preventing them from taking place.

The modes and mechanisms of failures caused by different defects in electrical connectors are discussed in detail in the following chapters dealing with different failure modes.

In general, there are several possible failure mechanisms corresponding to several possible defects or seed-defects which end up in the same *failure mode*. This will become clear from the following chart relating to roller bearing failures:

Terminology

Seed-defect or defect	Failure mechanism	Failure mode
Inadequate lubrication	Friction between races and rollers; overheating; scoring and chipping of active surfaces; more intense overheating.	Melting, fracture and welding of rollers and cages. (Bearing seizure)
Excessive interference between inner race and shaft	Cracking and slipping of inner race on shaft; fracture and chipping of rollers and races; intense overheating.	
Inadequate interference between outer race and housing	Creeping of outer race; wear; slipping of outer race in housing; overheating; scoring and chipping of active surfaces; more intense overheating.	
Excessive radial load on bearing	Premature fatigue cracks in races and rollers; chipping and spalling of active surfaces; overheating; intense overheating.	

It would be seen from the four examples given above, that the defects or seed-defects are different in each case. The mechanisms of failures are also different, but the final failure mode is the same. There are at least a dozen different defects and mechanisms of failures which end up with the same failure mode, i.e., bearing seizure.

It will be seen in the following chapters that there are many possible seed-defects and defects which have different failure mechanisms but the same end point: overheating, arcing, ignition of insulating materials and fire.

2.8 FAILURE RATES

When the numbers of similar equipment or components installed in any industry or group of industries are sufficiently large, i.e., in tens, hundreds or even more, it is useful to maintain statistics of failures and to calculate the failure rates periodically.

For example, if there are 200 switches of a particular make A, and 50 switches of another make B, it is possible to compare the reliability of the two groups of switches. If there were, say, 20 failures on switches of group A during a particular year, and if there were 12 failures on switches of group B during the same year, their FRs (failure rates) can be compared as follows:

$$\text{FRPCPY. } A = \frac{100 \times 20}{200 \times 1} = 10 \text{ percent per year}$$

$$\text{FRPCPY. } B = \frac{100 \times 12}{50 \times 1} = 24 \text{ percent per year}$$

The general formula for the failure rate is as follows (where FRCPY = failure rate, percent per year):

$$\text{FRCPY} = \frac{100 \times f}{P \times t}$$

where P = equipment population

f = number of failures in period t years

t = period under consideration, in years

(This formula is an approximation, but it is adequate for practical purposes).

It must be ensured, while making any comparisons of the reliability of equipment or components of different makes, that the failure statistics are considered for the same period. Further, if there are any differences in the conditions of service such as loading, environment or supply voltage, the failure rates would not be strictly comparable. There may also be differences in the failure modes. Despite such limitations, it is desirable to maintain statistics of failure rates of different types, makes and failure modes of electrical equipment and components. When wide variations are noticed between different groups, it is necessary to determine the reasons for the same. The reasons could be in specification, design, manufacture, operation, maintenance, environment, and service conditions. Such investigations often help to determine the ways in which failure rates can be reduced in any particular group showing a high failure rate.

If the number of hours that different equipment have worked per year are widely different, failure rates can be calculated on a per hour or per thousand hour basis, instead of per year basis. But then, the hourly utilisation statistics have to be maintained for all equipment.

2.9 INTER-RELATIONSHIPS BETWEEN TERMS

A clear understanding of the inter-relationships between the terms defined in the foregoing sections and illustrated by the flow chart in Fig 2.1 is essential for investigating fires and failures in electrical equipment and installations.

2.10 METAL FATIGUE

Whenever metal components like wires, hardware, and switch-gear components, are subjected to alternating (between tensile and compressive) or fluctuating stresses, they are vulnerable to a phenomenon known as *metal fatigue*. When the peak stress exceeds a certain limit (called the endurance limit), the metal component develops metal fatigue. The component develops cracks which grow in size with every new peak of the alternating stress. The rate of growth of the fatigue cracks depends directly on the excess of peak stress over the endurance limit and the number of stress peaks per second. Eventually, when the size of the fatigue crack becomes excessive, the component fractures. If this happens while the component is carrying a heavy current, an electric arc may be produced, and a fire may be the ultimate result.

Metal fatigue is different from human fatigue in many ways:
- A human being may get fatigued by merely carrying a heavy but steady load for a long time; but metals do not get fatigued by carrying such loads. They can sustain heavy loads for long periods without any ill effect, as long as the load is steady and within a certain limit known as the *yield point*.
- Alternating or fluctuating loads cause fatigue in metals even when they are much smaller than those that can be sustained in the steady mode; in humans such loads may even increase or develop the muscular strength.
- Humans can recover their strength and overcome the effects of fatigue by taking rest, but metals damaged by fatigue can never recover, either by rest, or by any process which is short of remelting or forging at high temperatures.

More about fatigue will be discussed in Chapter 13. At this stage, it may only be added that many failures and fires of electrical origin are caused by fatigue in metal components.

2.11 ELASTICITY OF METALS

Whenever a metal component is subjected to any mechanical load, it gets deformed in some way, such as by elongation, contraction, bending, or twisting. The nature of deformation depends on the manner in which the load is applied. If the load is within certain limits, the deformation is temporary. As soon as the load is removed, the deformation disappears completely and the component returns to its original shape and size. This property is known as the *elasticity of metals*. The load limit mentioned earlier depends on a property known as the *elastic limit of the metal*.

A very important property of metals, which is usually taken for granted, is that the elastic deformation of metals does not change with the passage of time. In other words, the elastic deformation which takes place initially as soon as the load is applied, remains constant for an indefinitely long time; and if at the end of that period, the load is removed, the metal component will return to its original shape and size, which it had acquired initially.

2.12 METAL CREEP

As stated above, the initial elastic deformation of metals remains constant for an indefinitely long time. This statement needs to be qualified by adding that this property is true only when the temperature is less than a value known as the *creep temperature limit*. This varies from metal to metal and depends on its melting point. The creep temperature limits (CTLs) for a few metals commonly used in electrical equipment are given below.

Metal	Creep Temp- erature Limit, °C	Remarks
Copper	135	In case of copper, creep can be a problem only in machines where conductor temperatures are high.
Aluminium	7	These three metals are subject to creep at normal operating temperature, i.e., 20 to 80°C.
Lead	− 70	
Tin	− 100	

Terminology

Above the CTL, metals are subject to a continuous increase in deformation as long as the deforming force is applied. The rate at which the deformation increases depends on the applied stress and on the rise in temperature above the CTL.

Exactly how and under what circumstances metal creep is responsible for failures of electrical equipment is explained in the relevant chapters of this book, and the phenomenon of metal creep is discussed in greater detail in Chapter 14.

2.13 STRESS AND STRAIN

The terms stress and strain are used rather loosely and interchangeably in ordinary conversations relating to human actions and reactions. In the field of mechanics, these two terms have totally different meanings which must be clearly understood.

The term stress means the reactive force per unit area produced inside a material which has been subjected to an external load or force. Thus, if a weight of 100 kg is supported by a wire of cross-sectional area 5 mm^2, the stress in the wire is calculated as $100/5 = 20$ kgf/mm^2. As the wire is under tension, this is a tensile stress. Similarly, if a weight of 1000 kg rests on a block of cross-sectional area 100 mm^2, the stress in the block is 10 kgf/mm^2. In this case, the stress is compressive [see Fig. 2.2(a) and 2.2(b)].

The term strain means the deformation per unit length produced in any material as a result of an external load. In the first example given above, if the original length was 10,000 mm and its elongation under the influence of the 100 kg load suspended from it was 10 mm, the 'strain' is $10/10000 = 0.001$.

Strain is defined as 'change in length' divided by 'original length'
(**Note** • Neither the force nor the cross-sectional area is involved in the calculation of strain.
• Strain has no unit. It is simply a ratio of two lengths measured in the same unit.)

The relationship between stress and strain may now be stated. An external load or force on a component produces an equal and opposite reactive force within the component. The force per unit area gives the stress in the material. The stress produces a strain in the material which is proportional to the stress. The strain multiplied by the length of the component gives the total elongation (or contraction) of the

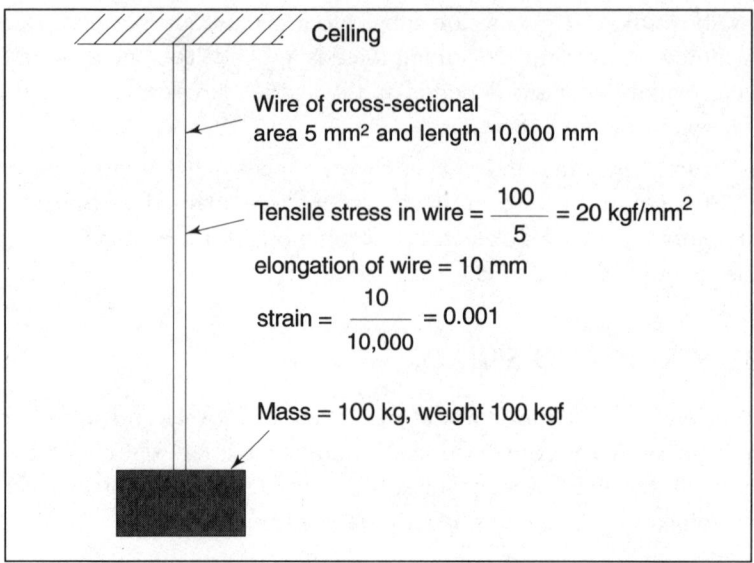

Fig. 2.2(a) Tensile stress and strain

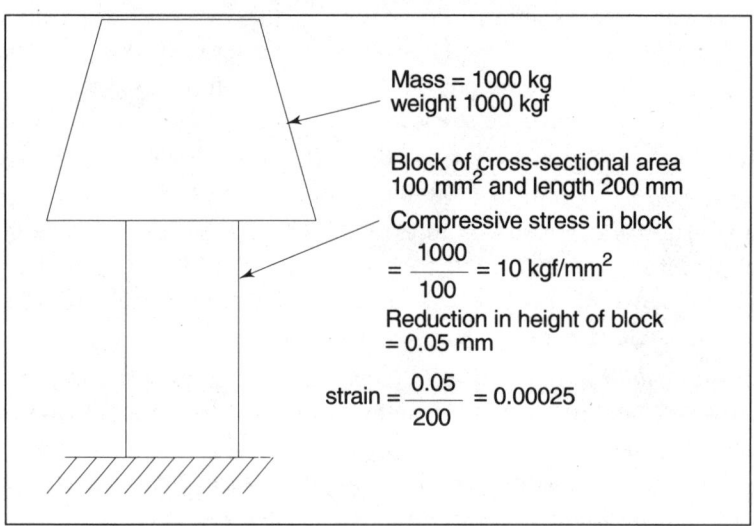

Fig. 2.2(b) Compressive stress and strain

component. The ratio of stress to strain is a constant which varies from metal to metal. For copper, it is about 12,000 kgf/mm².

This is a simplified example of a simple tensile or compressive stress on a component. More complicated examples of the effects of various

Terminology

types of forces on various shapes of components are studied by engineers under the subject 'Strength of Materials'. For our purpose here, it is sufficient to understand the difference between stress and strain.

2.14 ELASTIC LIMIT

In Section 2.11, it was stated that as long as the load is within a certain limit, the deformation is temporary and that as soon as the load is removed, the deformation disappears completely. This statement can now be restated since we have defined the terms stress and strain. As long as the stress is within a certain limit, the strain disappears completely when the stress is reduced to zero. This certain limit is known as the elastic limit of the material. It is a characteristic of the material. Different materials have different elastic limits. The elastic limit does not depend on the shape and size of the component.

From the definition of the elastic limit as given above, it should be clear that if the stress applied to a material exceeds its elastic limit, the deformation would not disappear completely on removing the stress. There would be some permanent residual deformation.

2.15 ENDURANCE LIMIT

In Section 2.10, the term endurance limit has been referred to as the peak value of an alternating stress limit above which metals get fatigued. The endurance limit is usually about 40 percent of the ultimate tensile strength (or the steady stress at which the metal will fracture) in case of ferrous metals. For non-ferrous metals, this limit is defined in terms of a certain number of cycles of stress (500 million) that can be sustained without fracture. Further details can be obtained from *The Metals Handbook* published by the American Society of Metals, but in this book, it is sufficient to remember that fatigue cracks and fractures are likely to occur if the peak stress exceeds the endurance limit—a limit which is much less than the ultimate tensile strength of the material.

It follows from the foregoing discussion that if the peak stress is less than the endurance limit, there is no possibility of fatigue cracks. Therefore, it is very important to ensure that the stresses on a component do not exceed this limit.

The term endurance limit will be discussed in greater detail in Chapter 13 on Metal Fatigue.

2.16 INSULATION RESISTANCE AND DIELECTRIC STRENGTH

These two terms are generally well known amongst electrical supervisors and engineers. Nevertheless, it is desirable to recall certain aspects. Both these properties of electrical installations and equipment are important, and to some extent, they are inter-related but they differ in their significance.

Insulation resistance is a measure of the leakage current across the entire insulation system and gives an indication of the general condition of the insulation. Insulation resistance is usually measured in Megohms. Some installations may show 'infinity' as the insulation resistance. This is not quite correct. It only means that the insulation resistance is higher than what the instrument is capable of measuring.

Dielectric strength is a measure of the minimum voltage at which the insulation would break down and develop a short circuit. It is indicated in kilovolts. It is not practicable to actually test this property on an installation because the damage is irreversible. The dielectric strength depends on the weakest point in the system. Even if the general strength is very high, one weak point will reduce the dielectric strength of the entire system, whereas there may be hardly any effect on the insulation resistance of the system. Therefore, high insulation resistance is necessary but not sufficient to ensure that there is no failure of insulation; the dielectric strength must also be satisfactory.

While the dielectric strength of a system cannot be measured, it is customary to proof-test it for one minute at a lower voltage known as the proof test voltage. The proof test voltage is generally approximately 40 percent of the dielectric (breakdown) strength as determined by design calculations and proved by destructive tests on prototypes or samples. The proof test voltage is approximately two to three times the maximum operating voltage.

It is desirable to verify both the above properties, viz. insulation resistance and proof test voltage for all new equipment and installations, before they are brought into use.

The well-known Megger test is a combined test for both the insulation resistance and dielectric strength. If a 1000-V Megger is used, the insulation resistance is measured and at the same time, a proof-test is carried out at 1000 V for one minute.

Chapter 3

Electrical Fires

In this chapter, we will discuss briefly:

🔌 The four main groups of causes of electrical fires:
- overloads • insulation failures • pressure contact failures
- conductor fractures

🔌 The importance of protective systems for the prevention of fires caused by overloads and insulation failures.

🔌 The non-availability of protective systems for the prevention of fires caused by pressure contact failures and conductor fractures.

🔌 The measures to be taken to prevent electrical fires due to all the four groups mentioned above.

🔌 3.1 INTRODUCTION

Fires which originate from defects in electrical equipment are called electrical fires. A few examples of such fires are briefly described below:
- The insulating board which formed the front panel of an electrical equipment cubicle for the control of a 50-hp electric motor got charred between two adjacent terminals. There was a short circuit and a momentary arc. The circuit breaker operated immediately and disconnected the power supply to the detective zone. The motor stopped and there was no fire. Repairs were effected and normal operations were restored within four hours (see Fig. 3.1).

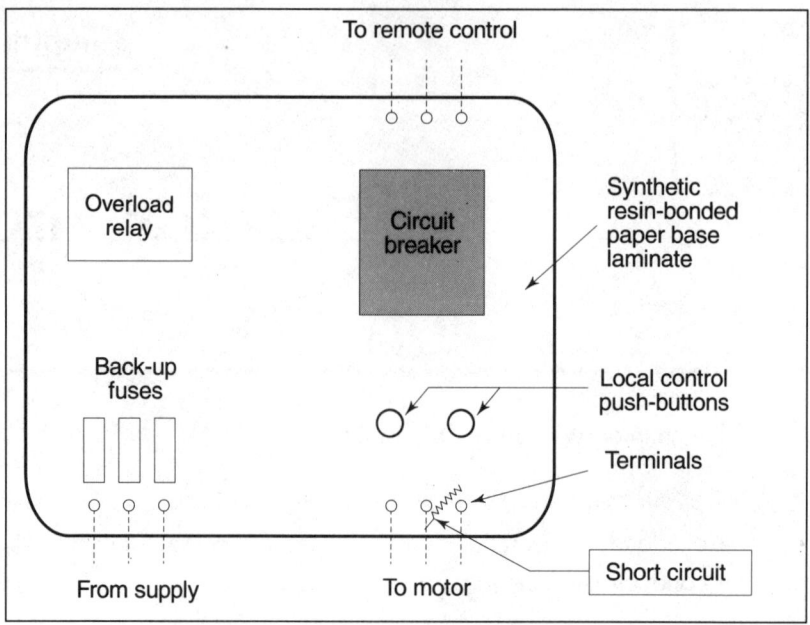

Fig. 3.1 Short circuit on a control panel

- In another incident, a similar failure occurred in an unattended area of a power station. The local protective system failed to operate. The arc continued to flare and the entire front panel caught fire. The equipment was totally destroyed before the fire was detected. The motor was also damaged due to single phasing. It took six weeks to obtain the required materials and to repair the cubicle and motor. Generation was restored after 18 hours of shutdown by installing a spare cubicle and motor.
- A parallel clamp which connected a lead-wire to an overhead transmission line overheated and melted. One of the three lines fell to the ground. There was arcing and some *jhuggies* caught fire. Three persons were electrocuted. It took six hours to carry out repairs and restore power supply to an industry (see Fig. 3.2).
- A transformer in an outdoor substation suddenly exploded. Supply to a residential area was restored after three days by replacing the burnt transformer. Investigation showed that parts of an off-load tap changer inside the transformer had melted down. The transformer windings and tank were badly dmaged. Repairs took several months.

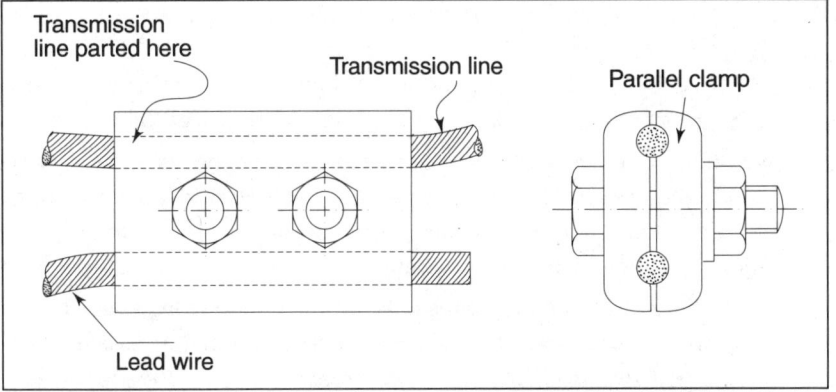

Fig. 3.2 Failure of a parallel clamp on an overhead transmission line

- A smoothing reactor in a 25-kV alternating current electric locomotive caught fire while the train was far away from any station. The train came to a stop and the fire spread to the interior of the locomotive. The fire extinguishers in the driving cabs were unable to control the fire. All the equipment in the locomotive was completely burnt down before any assistance could reach the site. The train was delayed by several hours and the locomotive was so badly damaged that it had to be written off from further service.
- In a pavilion, the temporary wiring was fixed close to the synthetic cloth used for the walls and roof. As the protective system was defective, a short circuit in the wiring led to overheating and continuous arcing. This caused the synthetic material to catch fire. Due to the inflammability of the material, and the wind direction, the fire spread very rapidly. In the stampede which followed, there was much loss of life.

There are many more incidents which are reported regularly in the press year after year. Moreover, there are hundreds of other cases of electrical fires which are not reported by the press either because they do not involve any loss of life or because they do not occur in public places. These are incidents which occur in factories, railway installations, power stations, warehouses, and buildings.

The so-called accidental fires are caused by a variety of reasons such as:

- Carelessly thrown lighted matches or cigarette butts

- Careless use of heating appliances like pressing irons and radiant heaters
- Careless storage and handling of highly inflammable materials
- Overheating, sparking or arcing due to electrical failures

We shall not discuss the first three of the four types of causes of fires mentioned above. We shall discuss fires of electrical origin starting in electrical equipment due to defects in the design, manufacture, installation or maintenance of the equipment.

The effects of electrical fires vary in magnitude from minor burning of an insulating board or an insulated cable to a major disaster involving several hundred fatalities and injuries amongst the public, apart from property losses worth crores of rupees. The severity of the effects depends on local conditions such as breeze, flammability of surrounding materials, availability and effectiveness of fire detection and fire fighting systems and the crowd behaviour under crisis conditions. All these are, for all practical purposes, beyond the control of the electrical engineer.

The best possible course of action to minimize losses is to attack the problem at its root, i.e. to prevent the fire from starting. It is possible to prevent electrical failures and zero failure performance of electrical equipment is well within our reach. The technology needed for this purpose is already available. The lacuna is only with regard to the training of the electrical technicians, supervisors and engineers. This book provides the course material for the required training. It should be possible to develop a three-day course from the material in this book.

3.2 SCOPE OF THIS CHAPTER

In Chapter 1, it was pointed out that seed-defects in electrical equipment and installations grow into defects and failures and that some failures develop into fires and, perhaps, disasters. In this chapter, it is proposed to discuss four main types of causes of electrical fires, viz.

(a) overloading
(b) insulation failures
(c) pressure contact failures
(d) conductor fractures

Whereas insulation failures and overloading are more frequent than pressure contact failures and conductor fractures, fires are more likely to be due to the latter two causes. The apparent paradox is explained by the fact that all electrical installations are provided with automatic protective devices which are capable of preventing fire when the fault is either overloading or an insulation failure [see Figs 3.3(a) and 3.3(b)].

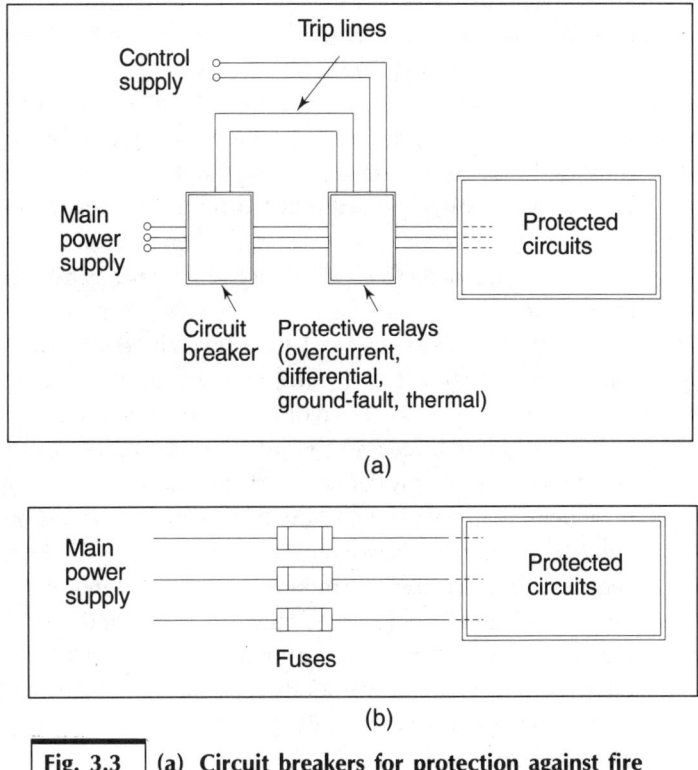

Fig. 3.3   (a) **Circuit breakers for protection against fire**
(b) **Fuses for protection against fire**

Unfortunately, there is no such device available for general use to guard against pressure contact failures or conductor fractures.

The prevention of electrical fires requires twofold action:
- Protection systems must be designed, manufactured, installed and maintained correctly.
- Great care must be exercised in designing, manufacturing and installing all components involving pressure contacts and conductors.

The technology involved in both the above lines of action is extremely simple. It is well standardized and quality assurance measures are very practical. Unfortunately, even the few simple rules which must be followed are not fully understood by many people, including some who are responsible for ensuring safety. The main need is for the training of the concerned staff to explain the significance of the apparently trivial requirements which have to be ensured. This book is aimed at fulfilling that need.

The need for a detailed discussion of this subject and dissemination of information arises out of the following facts:

(a) Electricity is an important necessity of modern life. Electrical installations are found in every home, office and factory.

(b) Although electrical installations appear totally harmless in their normal state, they can, without warning and at any time of the day or night, release vast quantities of energy into small spaces when they are defective. White hot temperatures are reached in fractions of a second.

(c) The consequences of fires can sometimes be disastrous, specially when fire detection and control systems are inadequate. This is usually the case. A totally avoidable and simple defect can lead to loss of life, injuries and loss of property, quite out of proportion to the cost of preventing such disasters.

(d) In general, a great deal of attention is paid to questions of fire fighting, rescue and restoration aspects not only by investigating agencies, but also by the media. Too little attention is given to the root causes of such fires. Investigations and enquiries into fires do not usually lead to any real preventive action.

(e) There are many misconceptions regarding the causes of electrical fires. If no other causes can be identified, it seems to be the usual reaction to ascribe the fire to 'an electrical short-circuit in the wiring' and to leave it at that, as if it is an uncontrollable natural phenomenon.

(f) Even if the fire is indeed due to a defect in the electrical installation, a short circuit is the least likely cause. The true cause of fire is more likely to be either a defect in the protective system or a defect in an electrical connector or terminal board.

(g) Since there is a great deal of damage around the true fault center, it is usually not possible to distinguish between cause and effect when looking at failed or burnt components. Erroneous conclusions may be reached as a result of errors in interpreting the visible evidence in the debris of the fire.

In this chapter, the modes and mechanisms of failures which culminate in fires will be discussed, as also some of the methods that can be adopted to prevent the introduction of seed-defects and the formation of defects which are the root causes of such failures.

Electrical Fires

We shall also discuss the investigation of electrical fires briefly in this chapter and in greater detail in Chapter 19. This is important because, as stated earlier, it is difficult to distinguish cause from effect and special methods have to be adopted to arrive at the true causes of electrical failures which end up as fires.

3.3 CAUSES OF ELECTRICAL FIRES

As mentioned previously, there are four main types of causes of electrical fires. In order to analyze the problem fully, it is desirable at the outset to distinguish clearly between these four possible failure modes due to different classes of defects in electrical installations.

(a) Overloading of electrical equipment results in excessive currents. As the heat developed in the cables is proportional to the square of the current, they get overheated. The insulation on cables is generally made of materials which are damaged easily by excessive temperature. They may therefore lose their insulating properties and lead to short circuits. Since many insulating materials are combustible, they may even catch fire if the temperature rises to their ignition temperature.

(b) Defects in or deterioration of electrical insulation may result in short circuits and continuous arcing, followed by ignition of the combustible insulating materials. Electric arc temperatures are extremely high and combustible materials in the vicinity catch fire unless the arc is extinguished within a fraction of a second.

(c) Deterioration and failure of pressure contacts between various components in the electrical installation may result in sparking, localised overheating and burning of combustible insulating material. In such cases, there may be no arcing initially, but the overheating is sufficient to start a fire. Short circuits and arcing may occur later, but they would not be the original cause of the fire.

(d) Fracture of current conducting components due to mechanical stresses or strains may result in local arcing at the point of fracture. Electric arc temperatures are extremely high and combustible materials in the vicinity catch fire.

It would be desirable to clarify the difference between 'sparking' and 'arcing' at this stage.

Sparking is often an early stage in the process or mechanism of failure of pressure contacts. If there is poor contact (for any one of several possible reasons) there will certainly be overheating. If the contact is momentarily broken but re-established quickly, there is sparking. This alternate break-make process may continue for hours or even days. It may even stop for a while and restart. Flickering of electric lights in houses is often due to loose connections in outlet boxes or between branch-circuit conductors and overcurrent devices in the service panel.

Arcing occurs when there is a positive gap in the path of the current, either when insulation fails or when a conductor or joint fractures while carrying a current. The arc current literally jumps across an air gap. The air gets ionised and the arc is maintained until the power is switched off by a protective device. Arc temperatures are of the order of several thousand degree Celsius. At such temperatures, the radiant heat transfer is often sufficiently high to ignite combustible materials some distance away from the arc.

The essential differences between the four modes of failures described above are shown in Figs 3.4(a) to 3.4(d).

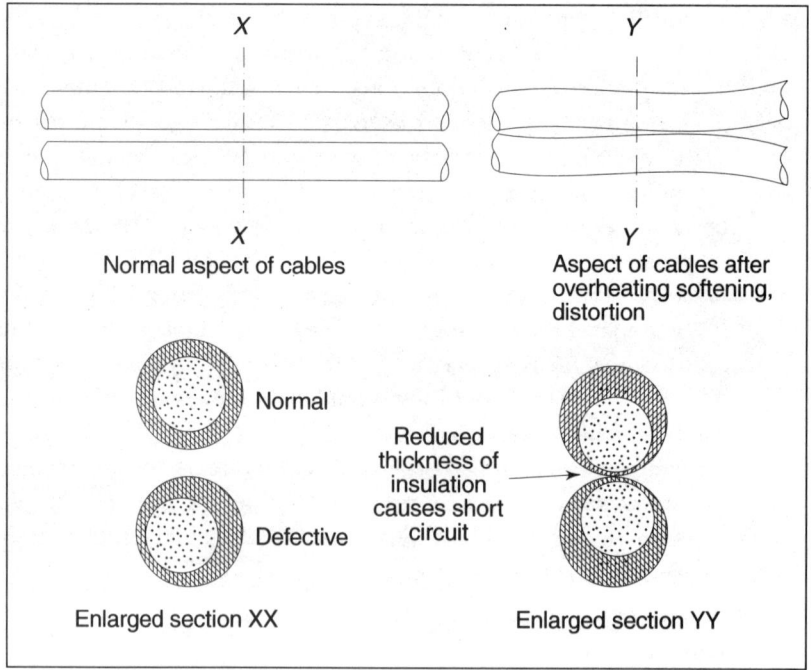

Fig. 3.4(a) **Overheating of cables due to overloading (in such cases, the entire length of the overloaded cables and associated fitting get overheated and distorted)**

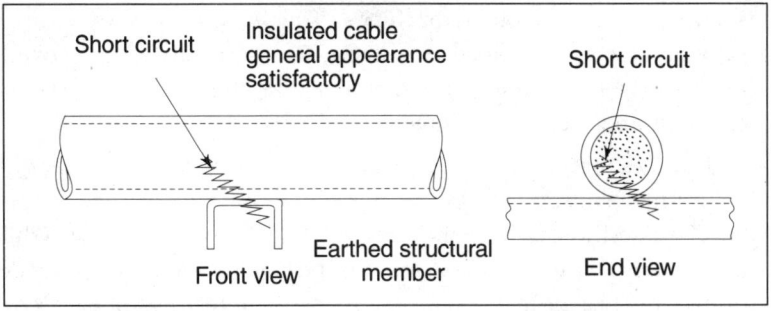

Fig. 3.4(b) Short circuit due to failure of insulation

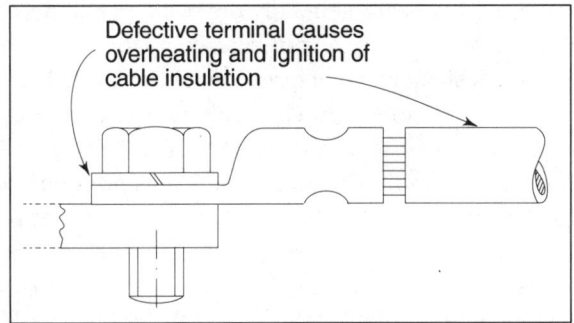

Fig. 3.4(c) Failure of pressure contacts (in these cases, there is local overheating of the cables near the failed pressure contact)

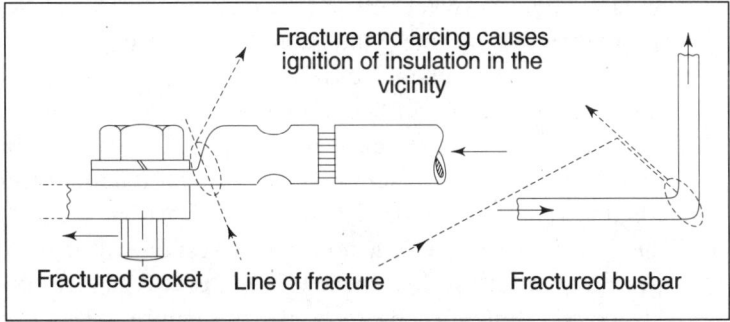

Fig. 3.4(d) Fracture of current carrying components followed by arcing across the fracture

Overloading of electrical equipment is usually the result of defective system design or unplanned or unauthorised additions to installations. The remedy is obvious. Sometimes, however, overloading of electrical machines may occur due to defects in the driven machinery such as pumps or other machines. The remedy in such cases is to provide adequate protective systems. Electrical fires due to overloading of driven machines are rare because the problem usually gets detected and attended in time.

The last three failure modes (b, c, and d) and the details of their failure mechanisms, as also the preventive or remedial measures, will be discussed in the relevant Chapters 5 to 12 which follow. While all these would, in the ultimate analysis, be due to defects in either design or manufacture or maintenance, a few special features of each of these three failure modes may be referred to in this preliminary discussion of the subject.

(a) Insulation failures are generally due to degradation of the insulation either as a result of normal ageing or as a result of some defect in design, manufacture or maintenance. It is possible to minimize insulation failures resulting from normal ageing by timely, preventive replacement of cables; but the cables do not age at every location at the same rate and some failures may occur before the scheduled replacement is accomplished. Very few insulating materials are immune to such degradation, and there is no totally dependable, practical method available to monitor the condition of the insulation continuously. Insulation failures can be minimised but not totally eliminated. Prevention of electrical fires depends, therefore, on protective systems which can detect insulation failures, and automatically and instantaneously switch off the power supply to the defective zone, thereby preventing the ignition of insulating materials. The measures to be taken for minimizing insulation failures are discussed in Chapters 10, 17 and 18.

(b) Failures of pressure contacts are largely due to the degradation of the mechanical contact pressure due to a number of physical phenomena like thermal expansion/contraction, creep, shrinkage, wear, elastic deformation, and vibration. All these can be prevented from causing failures by the proper choice of materials, correct design to ensure proper mechanical stress levels and adequate care during manufacture and maintenance to ensure that essential design criteria are respected. The details of these are discussed in Chapters 5, 6, 11 and 15.

A very common type of pressure contact failure is seen when aluminium conductors are used with screw type terminals, particularly if the screw is pointed. If the screw is not tightened adequately, the initial contact is bad and there is overheating. On the other hand, if the screw is overtightened, the conductor gets deformed and dented due to the softness of aluminium. The conductor may even fracture within a short time after commissioning. As explained elsewhere, the phenomenon of creep relaxation actually loosens the contact even when the screw is tightened correctly and remains unmoved. One way or another, eventual failure is certain when aluminium conductors are used with screwed terminals.

Failures of aluminium conductors at the terminals can be minimized by the use of clamp-type terminals, in which a small clamp-plate is interposed between the screw and the wire. This helps to prevent deformation of the wire initially, but the problem of creep remains. Use of crimped sockets to terminate stranded aluminium conductors is also a practice which helps to prevent damage to the conductors during tightening of the screw. Care must, however, be taken in the selection and use of the crimping tools and sockets, as discussed in Chapter 8.

(c) Fractures of conductors are generally due to vibration, excessive mechanical stresses, and sometimes, due to the use of substandard materials. All these can be totally prevented by proper design and care during manufacture and maintenance. The likely defects and the precautions that need to be taken to prevent such problems are discussed in Chapters 5, 8, 9 and 13.

The number of types of defects which can lead to failures and fires is very large, but fortunately, every one of them can be prevented by taking a few simple precautions during design, manufacture or maintenance.

We may now discuss in greater detail the four failure modes referred to in Section 3.3.

3.4 FIRES DUE TO FAILURE OF INSULATION

In the popular press and even in some technical investigation reports, the cause of fire is often described as 'an electrical short circuit'. All electrical installations have two or three conductors which are insulated from each other and from earthed structures. The voltage between such conductors is usually the system voltage, e.g. 240 V, 480 V, etc. If there is any failure of insulation at any point, a short circuit is the result. This causes excessive current, overheating and arcing at the point of short circuit.

Failure of insulation is quite common and by no means infrequent in an electrical installation. Insulation failures can and do occur due to a variety of reasons: the insulation may have deteriorated, as all insulating materials are bound to do, due to the normal ageing process. Whereas metals can retain their mechanical properties almost indefinitely, there are very few insulating materials which retain their insulating properties forever. Unfortunately, these few insulating materials are brittle, rigid and hence unsuitable for cable insulation. Further, insulating materials are easily damaged due to abrasion, environmental effects, mechanical damage, overheating, and rodent attack. Insulation failures and short circuits occur quite frequently and regularly in any large installation.

Although electrical insulation failures and short circuits are not unusual and are to be expected to occur at any time without prior notice or warning, fires are not to be considered as the inevitable consequences of short circuits. In fact, every electrical installation is supposed to be provided with a system which prevents fires in spite of insulation failures and short circuits. Such systems come into operation only when there are short circuits; at other times, they remain dormant. In brief, therefore, while short circuits are to be expected, fires are not.

3.5 TYPES AND CAUSES OF INSULATION FAILURES

There are three main types of insulation failures:
(a) insulation failures due to initial defect
(b) insulation failures due to normal ageing
(c) insulation failures due to external damage

Insulation Failure Due to Initial Defect

Insulation failure due to initial defect or deficiency in the strength of the insulation may be caused by an error in either the design, manufacture or the installation of the equipment. This type of defect can be prevented by taking the following measures:
(a) using material with NRTL* certification, and manufactured by reputed manufacturers.
(b) getting the installation designed and installed by licensed electrical contractors.

---

* NRTL=Nationally Recognized Testing Laboratory.

Electrical Fires

(c) carrying out high-potential (hipot) tests on the ionstallation before energizing it.

The reason for carrying out a hipot test may be clarified here. Despite using materials which comply with industry standards and despite getting the work done by licensed or qualified contractors, there is still a small probability of some defect remaining in the installation. To guard against even this small chance, it is a legally mandatory requirement that the installation be subjected to a high voltage test before being energized. In this test, a high voltage which is two to three times the normal operating voltage is applied for one minute between the different main lines and also between the main lines connected together and the earth. The exact value of the test voltage is given in the applicable industry standards.

If there is any defect or weakness in the insulation, a fault will be indicated by the testing equipment. The fault will then have to be located, repaired and then the installation retested.

Once this hipot test is done satisfactorily, there should be no further work on the installation. If any work of repair, modification or extension is done, the test has to be repeated.

If the three precautions mentioned above are taken, the possibility of an insulation failure occurring in service will practically be nil and hence, there will be no fires on new installations.

Insulation Failures Due to Normal Ageing or External Damage

If certain precautions are taken, there is no possibility of any insulation failure or fire in a new installation. However, we have to guard against the effects of time and usage. Firstly, all insulating materials deteriorate in their insulating properties with the passage of time. Secondly, the rate of deterioration depends on many factors which are beyond measurement or control. Some of these factors are: the electrical loading on the wires, the ambient temperature, the layout and surroundings of the wires, presence of foreign materials, the original quality of the design and of the materials.

Therefore, despite the original installation being of the best possible standard, we must be prepared for insulation failures taking place in service after the installations have been in use for some time. This is done by providing what are known as protective systems.

Failures of electrical equipment due to insulation failures need not necessarily culminate in fires. If an automatic protective system is provided to detect the electrical insulation failure and to switch off the power supply instantaneously, there would be only some local damage, but there would be no fire. Such protective systems are indeed available at present. They consist, generally, of one or more of the following devices:
- Fuses
- Circuit breakers

Protective systems are usually provided in all electrical installations. In fact, it is necessary under the law to provide these safety devices. Why then do electrical fires take place? There are two reasons for this:
- Sometimes, the protective systems are not designed or adjusted correctly.
- In some cases, the protective system may have become defective in service and remained in that state due to inadequate maintenance.

The absence of or the defective state of a protective system constitutes a seed-defect. This will not lead to any fire immediately, but if and when a short circuit occurs anywhere in the unprotected zone, there is every possibility of continued arcing, overheating and ignition of combustible insulating material. Whether the fire spreads and results in a disaster or not, is now a matter of chance.

3.6 THE IMPORTANCE OF ELECTRICAL INSTALLATION CODES

The National Electrical Code® (NEC) published by the National Fire Protection Association in Quincy, Massachusetts, is the basis for official wiring safety rules in most U.S. states, cities, and counties. It has also been adopted in Mexico and is being considered for use in other countries around the world.

The NEC's stated purpose is "the practical safeguarding of persons and property from hazards arising from the use of electricity" (Section 90-1). In order to avoid becoming obsolete, it is revised and updated every three years to reflect new products and electrical construction technology.

Important excerpts from the National Electrical Code follow. Article 110 describes general requirements applying to all electrical installations;

important excerpts follow. *(Italicized material in parentheses is explanatory in nature and does not form a part of the official NEC language.)*

Article 110   Requirements for Electrical Installations

110-2. Approval. The conductors and equipment required or permitted by this Code shall be acceptable only if approved. *(By the authority having jurisdiction, which is usually the electrical inspector.)*

110-3(b). Installation and Use. Listed or labeled equipment shall be installed and used in accordance with any instructions included in the listing or labeling. *(Product certification agencies such as Underwriters Laboratories Inc. establish conditions for safe use of electrical products, based on NEC requirements.)*

110-7. Insulation Integrity. Completed wiring installations shall be free from short circuits and from grounds other than as required or permitted in Article 250. *(Article 250 covers grounding of electrical circuits and equipment for safety.)*

110-8. Wiring Methods. Only wiring methods recognized as suitable are included in this Code. These recognized methods of wiring shall be permitted to be installed in any type of building or occupancy, except as otherwise proved in this Code. *("Wiring methods" refers to conductors and cables, along with methods of support and protection such as raceways and cable trays.)*

110-9. Interrupting Rating. Equipment intended to interrupt current at fault levels have an interrupting rating sufficient for the nominal circuit voltage and the current that is available at the line terminals of the equipment.

110-12. Mechanical Execution of Work. Electrical equipment shall be installed in a neat and workmanlike manner. *(To supplement the NEC, a series of National Electrical Installation Standards is published by the National Electrical Contractors Association in Bethesda, Maryland, to define what is meant by "neat and workmanlike" electrical construction.)*

3.7   IMPORTANCE OF PROTECTIVE SYSTEMS

Protective systems do operate correctly many more times than they fail to do so; but this is not enough. They must operate correctly every time. Installations will continue to operate normally even when the protective systems are defective, but they will be vulnerable to incidents of fire. Defects which develop in the protective systems will remain undetected, unless fail-safe devices are used or periodical checks are organized. Higher levels of management have a special responsibility to introduce suitable checks to ensure that protective systems are invariably and correctly installed and that they are always in good working order.

If and when an insulation failure occurs due to any reason, be it original deficiency or normal ageing or external damage, the protective system should automatically and immediately cut off the power supply to the defective zone of the installation, thereby preventing any ignition of the insulating material. The heat developed in the short time taken by the protection system to operate is too small to raise the temperature to the level needed for igniting the material. Thus, although there would be an interruption in the service, there would be no fire.

This system of providing automatic devices to switch off power in the event of insulation failures, forms the very basis of safety for all electrical installations. Such devices are invariably provided to protect every circuit in all installations in domestic, commercial and industrial sectors.

Insulation failures can occur at any time and without any prior warning. There may be no visible or even measurable indication of impending failure. Therefore, the only protection against fires as a consequence of insulation failure, is an automatic protective system. Fuses, over current relays, differential relays and circuit breakers are some of the usual protective gear in use.

It is necessary not only to provide a system which is capable of detecting the fault currents which flow through the faulty section after insulation failures, but also to interrupt the very high fault currents automatically and almost instantaneously. The faulty sections must be isolated from the source of power.

It is further necessary to ensure, by using reliable protective systems and by making periodic tests, that the system is maintained in good working condition. If the protective system is allowed to remain inoperative while the main installation is in service, there is always the danger of an insulation failure and short circuit occurring at that time. In that case, a fire is almost certain to flare up. If there is delay in the detection and extinction of the fire, the incident can turn into a major disaster.

In brief, therefore,
- Insulation failures can be minimized by periodic condition monitoring of the insulation, but they cannot be completely eliminated.

- Despite the occurrence of insulation failures, fires can be totally prevented by:
  (a) provision of an effective protection system, and
  (b) maintenance of that protective system in good condition at all times

It is the protective systems which make all the difference between safety and fires. The entire electrical system (generation, transmission, distribution and utilization of electricity) is completely dependent for its survival on the performance of the protective system. Thousands of short circuits occur every day, but very few of these result in fires. In these exceptional cases, there is usually some defect in the protective system. Insulation failures can be minimized, but cannot be prevented; but electrical fires resulting from short circuits are certainly preventable.

Article 240 of the National Electrical Code requires overcurrent protection for all branch circuits and feeders. Molded-case circuit breakers (MCCB) and fuses are normally used for circuits operating at typical building voltages (120 to 480 V), while draw-out type circuit breakers are often used to protect high-voltage circuits.

If a correctly designed protective system is provided, it is impossible for a fire to be started as a result of a short circuit.

If the protective system is defective, a fire is bound to occur sooner or later.

Short circuits cannot be totally prevented from occurring, but fires can be prevented by providing effective protective systems.

Therefore, short circuits cannot be considered to be the causes of fires. The real causes are defects in protective systems.

Figure 3.5 shows the schematic arrangement of a simple protective system, a block diagram for which is shown in Fig. 3.3(a). Every protective system consists of two major elements: (a) the fault detection system comprising one or more detection devices, and (b) the fault isolation system comprising circuit breakers. These two are interconnected and the combined operating time of both is usually a fraction of a second.

If a correctly designed and well-maintained protective system is in operation, all that can happen when there is a short circuit for any reason whatsoever, is that there would be an interruption in power supply to the user device. The faulty section of the installation would

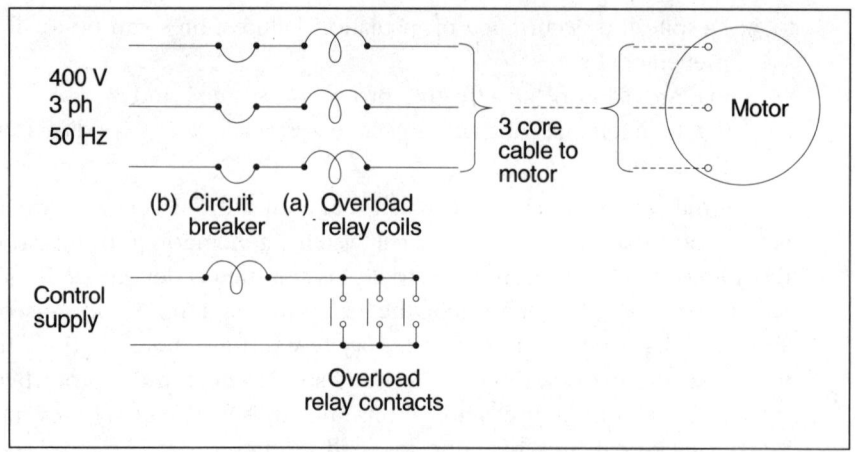

Fig. 3.5 Schematic diagram of simplified protection circuit for motor and cable

get disconnected from the source of power within a few hundredths of a second. There would be only a slight overheating at the point of short circuit. Often, the point of short circuit would not even be visible and it may take some effort to locate it. The heat generated at the point of short circuit would be quite inadequate to ignite any ordinary combustible materials.

(In case of explosive or gaseous combustible materials around the equipment, fires may start even by the little arc or spark which is quenched quickly by the protective devices. However, these are very special cases in certain specified industries which are required to take further special precautions such as the use of flame-proof equipment. With these additional measures, accidents can be prevented from occurring even in such hazardous locations.)

Protective systems tend to be neglected because they take no part in the normal operation of the equipment. Production or service can continue even when the protective system is inoperative. There are even cases where protective systems are either ignorantly or deliberately by-passed or modified. A common example of this practice is the replacement of the correct fuse by a fuse of a higher rating in an overloaded circuit. Some protective systems utilise precision relays and these have to be correctly adjusted and calibrated. They must also be periodically verified for correct operation.

Electrical Fires

Some fires of electrical origin may be due to design, manufacturing and maintenance defects in the protective systems. The design and testing of all types of protective devices and the measures to be taken to prevent defects in the protective devices are beyond the scope of this book. A few important facts about such devices used in low voltage, low power installations may however be discussed briefly. Many fires in residences and commercial buildings and even in industrial installations are due to defects in low voltage, low power electrical systems.

3.8 PROTECTIVE SYSTEMS—FUSES

The simplest and most common protective system is a fuse. Fuses can be used in installations operating at 240 and 480 V. Fuses can be used at even higher voltages, but it is now customary to use circuit breakers for higher voltages and currents.

A fuse consists essentially of a wire of cross-section smaller than the cross section of the wire to be protected. The cross section of the fuse wire is so selected as to ensure that the fuse wire would heat up and melt, thereby breaking the circuit, before the main wire being protected reaches too high a temperature.

The fuse wire is enclosed in a ceramic cartridge or fuse holder, so that there is no danger of starting a fire from the fuse itself.

Fuses are fail-safe. The effect of ageing is the reduction of the current at which the fuse will blow. There is a possibility then of the fuse blowing even under normal conditions, but there is no danger that the fuse may fail to blow under short circuit conditions.

If the fuse wire is made of silver or silver-plated copper, the condition of the fuse wire does not deteriorate even after years of use. The fuse wire is then not only fail safe, but also reliable. It will not blow under normal conditions.

It is important to use high interrupting capacity fuses where the fault current levels are high. If an ordinary fuse is used in such a location, the fuse will fail to clear the fault. It may explode and continue to arc and start a fire from the fuse board. It may be stated here that this kind of problem arises only in substations and power stations. This is an aspect which needs to be looked into by the design engineers in charge of system design.

3.9 PROTECTIVE SYSTEMS—CIRCUIT BREAKERS

A variety of circuit breakers are now available–miniature circuit breakers, and moulded case circuit breakers for low current systems; for high current, high voltage systems, it becomes necessary to use oil circuit breakers, vacuum circuit breakers, air blast circuit breakers, or SF6 circuit breakers. They are capable of disconnecting the faulty section from the source of power even while carrying very high fault currents.

Circuit breakers usually have complex and precision mechanisms. They may be operated either by hand or by means of electromagnetic or compressed air motors. The circuit breakers may be suitable for local or remote control. A large variety of complex protective systems such as overcurrent ground-fault and differential, are also in use.

Some circuit breakers are equipped with built-in overcurrent trip devices, and also thermal trip devices which can take care of normal short time overloads.

Circuit breakers have the advantage of being instantly reusable while fuses have to be replaced or rewired. However, circuit breakers have the disadvantage of not being fail-safe. They need to be checked and overhauled periodically. It is often advisable, particularly with miniature circuit breakers, to provide back-up fuse protection.

Miniature Circuit Breakers have the added limitation that they are not designed for maintenance, but they are vulnerable to the effects of dust, arc-erosion and sometimes corrosion. Therefore, it is necessary to install Miniature Circuit Breakers in dust proof enclosures and to test them periodically to check that they trip at the specified currents within the stipulated timings.

3.10 PREVENTION OF FAILURES OF PROTECTIVE SYSTEMS

Since the only practical way to prevent fires due to short circuits caused by insulation failures is to ensure that the protective systems do not fail in service, it is necessary to examine this aspect in greater detail.

Protective systems will include items like fuses, over-current relays, circuit breakers, and current or voltage transformers. The first requirement is that all such items included in the protective system must be intrinsically reliable and fail-safe. They should be obtained from manufacturers who can supply documentation in the form of test results to prove these characteristics. Fail-safe means that even in the rare event of a failure of the equipment, the failure should be on the side of safety. In other words, the equipment may operate when there is no fault, but it must not, under any circumstances, fail to operate in the presence of a fault in the protected zone.

After installation for the first time, the protective system should be tested by simulating faults of the type the system is supposed to guard against.

A maintenance system should be introduced to check periodically, the calibration and proper performance of the protective system and its components. Some organisations consider it prudent to establish a separate department—distinct from the routine maintenance department—for the design, testing, calibration and monitoring of all protective gear. This is particularly true in the case of relays and circuit breakers. Fuses are intrinsically fail-safe as long as their original selection or installation is done correctly.

3.11 FIRES DUE TO FAILURES OF PRESSURE CONTACTS AND FRACTURES OF CONDUCTORS

Out of the three classes of defects which lead to fires, we have so far discussed only the first, i.e., fires due to failure of insulation. We will now discuss fires due to the other two defects—failures of pressure contacts and fractures of conductors. These two types of defects are discussed together because they have one common feature:

There are no protective systems available which can automatically and immediately disconnect the power supply when failures of pressure contacts or fractures of conductors occur in service.

In other words, there is no second line of defence against fires arising out of such defects.

The protective systems which are usually provided are only capable of detecting short circuits or earth faults caused by insulation failures. There are no protective systems available for detecting failures of pressure contacts or fractures of conductors. When such defects occur, overheating and arcing will take place and continue to grow in intensity until a fire is started. It is therefore of utmost importance to ensure during design, manufacture, installation and maintenance, that neither defects nor seed-defects which could develop into such failures of pressure contacts or fractures of conductors are introduced at any stage.

Sometimes an insulation failure and a short circuit may occur as a result of the heat produced by the failure of the pressure contact or fracture of a conductor, and this may cause the protective system to operate. In such cases, the short circuit is a blessing in disguise, but we cannot be certain that this would always happen before a fire is ignited.

The mechanisms of failure are totally different in the above two cases and it is necessary to understand them very clearly, because, while the fire may originate due to only one of the two types of defects, defects of both types may be observed after the fire. The distinction between the different failure mechanisms are shown in the flowchart in Fig. 3.6, which shows how an electrical fire may start.

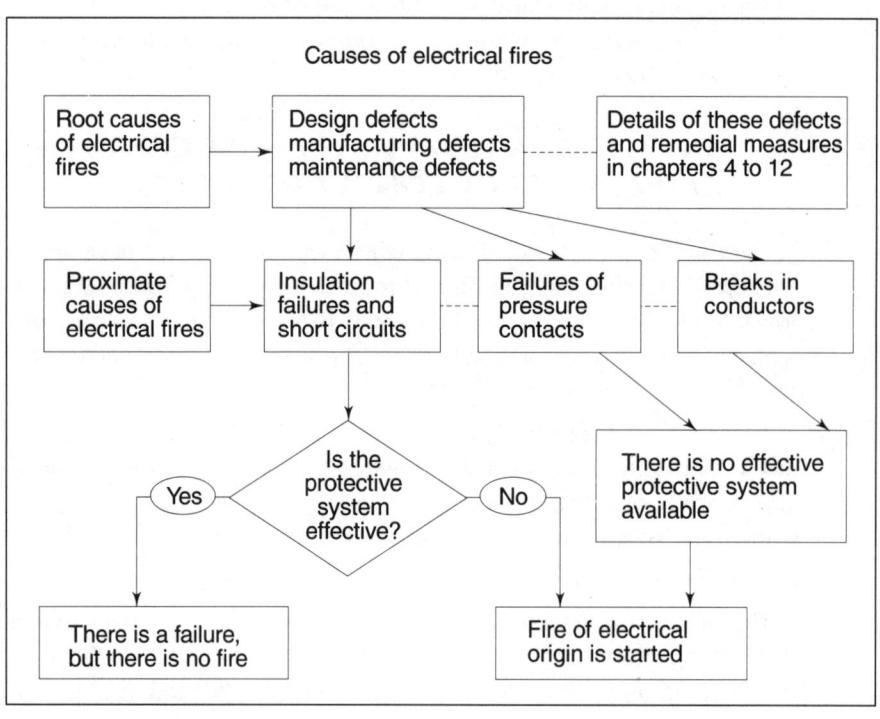

Fig. 3.6

There are thousands of points in a large electrical installation where electric current is transferred from one component to another through a pressure contact in which the pressure is developed by threaded

fasteners, springs or elastic deformation. Some examples of such points are:
- cable strands and crimped socket
- crimped socket and terminal
- terminal and busbar
- busbar and equipment terminal
- two busbars

At all such points there is additional resistance, and consequently, some additional generation of heat. The resistance and the temperature of the components is usually low. In general, the temperature rise is less than 40°C. However, under certain circumstances, the contract resistance can get out of control. The contact resistance depends on the contact force, and if there is any loss of this contact force for any reason, there is overheating at that point which can get out of control, and high temperatures, at which nearby insulating materials can ignite, may certainly be attained.

Fractures of cable strands and crimped sockets can also lead to electrical fires, because if the final fracture or parting takes place when a heavy current is passing through the cable or socket, an arc is produced which may continue to flare for quite some time (unlike arcs caused by short circuits which are usually quenched in a few hundredths of a second by the protective devices).

Defects in connectors and terminal boards can therefore result in electrical fires, because there is no protective system now is general use which could detect the development of hot spots or arcs due to such failures. The protective systems provided for detecting short circuits may sometimes prove useful if a short circuit or ground fault also develops as a result of damage to the adjacent insulation in the early stages of the arc or overheating caused by a failure of an electrical connection. In such cases, a short circuit is a blessing in disguise if it occurs soon enough to prevent a fire.

As there are many different types of defects which can and do occur in electrical connectors, these are discussed in greater detail in Chapters 5 to 12.

3.12 FIRE PREVENTION MEASURES

The measures to be taken to prevent defects in and failures of electrical connectors and terminal boards will be discussed in detail in Chapters 5 to 12. It will be seen that the measures suggested are simple and practical. They will certainly prevent fires and, in addition, a number of failures will also be prevented from occurring. The costs involved in these measures are negligible; the benefits in terms of savings in repair/rehabilitation costs and prevention of revenue losses would be very significant.

As mentioned above, there are no protective devices now in general use which can detect, in an incipient stage, the development of hot-spots in any one of the thousands of points at which connector or terminal failures can occur. Special care must therefore be taken on the lines suggested in the chapters mentioned above.

There is, however, one hot-spot monitoring device which is now available. It is not in general use mainly because of its high cost. This device is the infrared video camera with a CRT monitor. This can be periodically used to inspect all the vulnerable connectors while in service to detect hot-spots. As it is not necessary to come close to the equipment, it can even be used to inspect high-voltage installations. Since the degradation of connector joints is very gradual and takes quite some time, it is possible to detect developing defects well in time to be able to attend to them, thereby preventing a certain failure and probable fire. This device, as also similar devices, are discussed in Chapter 11.

While it may not be possible, due to the high initial cost, for the owners of small installations to invest in such monitoring equipment, large industries and organizations can certainly do so. Perhaps there is an opportunity here for small entrepreneurs to provide such monitoring services for a reasonable fee. One way or another, all electrical installations of consequence should be periodically subjected to infrared examination. This will go a long way towards the elimination of electrical fires.

3.13 CONCLUSION

Electrical fires are more likely to be due to the failures of electrical connectors and terminal boards, because there are no protective systems to detect bad contacts. Failures due to short circuits are less likely, because protective systems are usually provided, and if a fire still occurs, it would be due to a defect in the protective system.

The measures or precautions to be taken to prevent failures of electrical connectors and terminals are very simple, inexpensive and practical. It must be emphasized that it is not sufficient if foremen and supervisors are knowledgeable in these matters. It is the staff who actually do the work with their own hands, who should be trained about the correct and incorrect practices. Moreover it is also not sufficient if the majority of such staff are well trained. It is necessary to ensure that 100 percent of the staff who actually do any work are trained fully to do that work correctly. Electrical connectors are all fitted manually and it is usually not practicable to test or inspect each and every one of the thousands of locations in each installation. It is necessary that the staff who do the work do it right every time.

3.14 DO'S AND DON'TS FOR PREVENTING ELECTRICAL FIRES

- Ensure that current ratings of wires, cables and accessories are equal to or higher than the maximum possible loads.
- Ensure that all wires and cables are protected by fuses.
- Ensure by using branch fuses that every bit of wire or cable is protected by a fuse of equal or lower rating.
- If MCBs or cartridge fuses are used in place of wired fuses, test them in accordance with relevant Indian Standards before acceptance.
- If MCBs are used, provide back-up fuses.
- With regard to connectors and terminals, follow the Do's and Don'ts given at the ends of Chapters 5 to 12.
- Ensure that switchboards, terminal boards, joints, etc., are not fixed close to combustible materials.
- In high current and high voltage installations where circuit-breakers are provided, ensure that protection systems are calibrated and adjusted periodically as recommended by the manufacturers.
- In the absence of infrared cameras, carry out at least close visual monitoring of all bolted or crimped joints. Look for tell-tale signs of overheating, such as softened, distorted or charred cable insulation or terminal boards. (*Caution*: This can be done only after switching off the power and earthing the circuit being checked.)
- Use, where possible, infrared cameras and video monitors to identify connectors and terminals in high current installations which are operating at temperatures above 85°C.

- Use only listed electrical equipment, cables/wires and accessories.
- Investigate fully every case of electrical fire and all cases of repetitive failures to determine the root causes and preventive measures.

Chapter 4

Transformer Failures

In this chapter, we shall discuss:

- Failures of oil filled power transformers of the type used in power stations, grid sub-stations, and local area sub-stations.

- The magnitude of the problem of transformer failures. Out of all the different types of electrical equipment in power stations and distribution networks, only the transformer has been singled out for discussion in this book for two reasons:
  - Transformer failures often result in explosions and fires in public places.
  - Most of the transformer failures are due to mechanisms of failures which have been discussed in detail in this book.

- The common failure modes of power transformers. Other less common failure modes are also enumerated.

- The mechanisms of failure and the preventive measures which can be taken in regard to each of the common failure modes.

4.1 INTRODUCTION

Transformers are electromagnetic devices without moving parts that change values of voltage, current, and impedance without changing frequency. Basic transformer construction consists of a soft iron core with two wrappings or coils of wire. The coil on the line or input side is known as the primary. The coil on the load or output side is called the secondary.

Transformers range in size from miniature components used in electronic devices to large devices rated in hundreds or even thousands of kilovolt-amperes (kVA) used in utility and industrial power systems.

From the power stations at which electriciy is generated up to the points where it is utilized in residential, rual, commercial and industrial units, electric power goes through several voltage conversions by means of power transformers. These repeated voltage conversions become necessary because the most economical voltages for different purposes are different and generally as follows:

Application	Suitable voltage (kV)
Generation	6.6 or 11
Long distance transmission	400
Medium distance transmission	66 or 132
Short distance transmission	11 or 33
Distribution and utilisation	0.4 or 3.3

Figure 4.1 shows a typical electrical network system in which power is transformed repeatedly to voltages most suitable for the different parts of the system.

All the electrical power that is generated in power stations is repeatedly transformed from one voltage to another, and the total capacity of installed transformers is approximately five times the installed generating capacity.

Power transformers are a critical component of utility distribution systems. Their individual capacities vary from a few hundred kVA to several hundred MVA. The reliability of distribution transformers of low capacity is poor, mainly due to defects and deficiencies in the design of minor components and process details. The cost of the transformer depends mainly on the design of the core, the coils and the tank. The designs of these major subassemblies are generally satisfactory. The defects are in the design of the small hardware, the cost of which is very small in relation to the total cost of the transformer.

These power transformers are static units with no moving or wearing parts. If they are properly designed, manufactured and maintained, they should operate with a failure rate which is less than 0.1 percent per year. Actual failure rates in different organizations are in the range 2 to 20 percent per year (i.e., 20 to 200 times the optimum failure rates). There are a few organizations in India which have achieved failure rates as low as those in developed countries, but they are exceptions. They prove that it is indeed possible to achieve very high reliability of transformers.

There are many reasons for the high failure rates of transformers in most organizations. The failures are mainly due to avoidable defects or

Transformer Failures

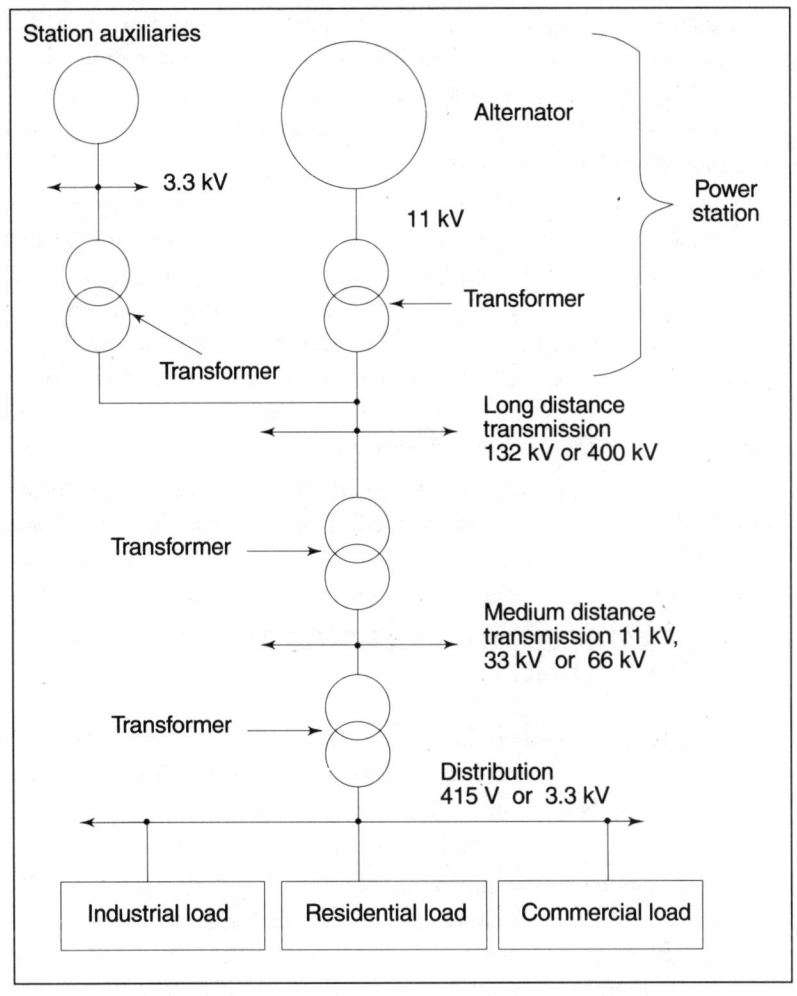

Fig. 4.1 Typical electrical power network

deficiencies in design, manufacture and maintenance. It is estimated that about 50 percent of the failures are due to defects in connectors and terminals, the remaining being due to all other causes put together.

When transformer failures do take place, it is generally felt that these are inevitable. The usual reasons which are trotted out are either overloading or natural phenomena such as heavy rain or lightning. The fact that transformers can work reliably despite these conditions is not appreciated. It is the purpose of this chapter to explain the proposition that zero failure performance can be achieved, with the technology and materials available in India. The few simple precautions that need to be taken in the design, manufacture and maintenance of transformers will be described and discussed in the following paragraphs and chapters.

Electrical connectors and terminal boards are used inside almost all electrical equipment, and the transformer is no exception. As in the case of many other equipment, a significant proportion of the total number of failures of transformers are due to failures of electrical connections and terminals. The failure modes, failure mechanisms and the preventive measures relating to these apparently 'minor' components will be discussed in Chapters 5 to 12. In this chapter, the other common modes of transformer failures are discussed in detail.

Certain general precautions to be taken at the stage of specification, detailed design review and inspection of the prototype will also be discussed in this chapter.

4.2 ARE TRANSFORMER FAILURES DUE TO OVERLOADING ?

Some electrical engineers in charge of transformer maintenance, often put overloading as the cause of failures of their transformers. Many of them sincerely believe that this is the true cause of the failure, and do not pursue further investigations to identify other possible causes.

While it is true that prolonged overloading beyond certain limits can indeed lead to premature failures of transformers, it must be noted that transformers are intrinsically very rugged devices. They can withstand considerable overload without permanent damage. It is necessary, therefore, to understand clearly the permissible limits of overloading.

Actually, it is also a fact that transformers do get overloaded from time to time and failures do sometimes occur when the transformer is operating with an overload. However, detailed investigations often show that overloading is only a proximate cause and that the root cause of the failure is altogether different. This becomes clear on examination of the transformer specifications and the 24-hour load curve.

Transformers which comply with industry standards have considerable in-built overload capacity. For instance, a 50 percent overload may very safely be imposed on a distribution transformer with natural oil and air cooling, for periods as shown in the following table:

Loading in remaining period (%)	50% overload permissible for (hours)
90	1
60	2
25	3

The overload capacity referred to in the preceding paragraph is very significant because the loading on most distribution transformers fluctuates greatly and there are long periods of underloading. Advantage can be taken of this feature during short peak-load periods within the permissible limits as above. If these limits are respected, there is no loss, whatsoever, in the reliability and durability of the transformer.

Even if these permissible overload limits are transgressed occasionally, it does not follow that the transformer would fail very soon on that account. All that happens is that the durability of the transformer is reduced by a period which is directly proportional to the duration of the overload, and exponentially proportional to the excess of oil temperature above 90°C.

For every 6°C rise in the oil temperature above 90°C, the life of the transformer is reduced by a period which is double the period for which the transformer is used at that temperature. The critical factor which determines the loss of life is the temperature of the transformer oil. Normally, i.e., when the transformer is designed, manufactured and maintained correctly, the life of the transformer may be taken as about 40 years. For every hour of operation with an oil temperature which exceeds 90°C by the following amounts, the loss of life would be as shown:

Excess in oil temperature above 90°C	Loss of life for every hour of operation (Hours)
6	2
12	4
18	8
24	16
30	32

(Operation of transformers with oil temperatures in excess of 120°C or hot spot temperatures in excess of 140°C is not permissible.)

Thus, an occasional short time overload which causes the oil temperature to rise beyond the normal limit of 90°C up to even 120°C would not cause an immediate failure of the transformer. It may reduce the life of the transformer from 40 years to, say, 38 years, assuming worst case operation (i.e., operation at 120°C oil temperature) for one hour, once every month.

The foregoing discussion is certainly not meant to encourage systematic overloading of transformers. Every effort must be made to provide transformers of adequate rating to ensure that the top oil temperature is generally below 90°C, since, in the long run, persistent overloading is likely to be uneconomical.

The effects of overloading on transformer reliability and durability have been considered here only to emphasise that overloading should not be considered to be the cause of failures unless calculations based on detailed observations of load curves and oil temperatures show that the main conductor insulation is being severely damaged by high oil temperatures. A reference may be made to NEMA standards for guidance with regard to such calculations.

In most cases, it is usually possible by making a few simple observations and calculations, to determine that overloading is not the main cause of transformer failure, even though the failure may have occurred while operating on overload. It then becomes necessary to examine the other possible and more likely causes of failures as suggested in the following paragraphs.

4.3 FAILURE MODES OF POWER TRANSFORMERS

In most cases it is usually possible by making a few simple often not possible to distinguish between cause and effect of failure by merely looking at the burnt components of the transformer. It can be stated, moreover, that almost all the cases of transformer failures start as a short circuit from a very small point. However, the energy released at that point (within the short time between the occurrence of the short circuit and the disconnection of the transformer from the source of supply) is very large. It is large enough, in fact, to melt copper conductors and to char/ignite the insulating material. If there are effective protective devices, the damage may be small and the exact point of origin of the fault may be identifiable. If the protective device is ineffective or defective, much damage, including an explosion and an oil-fire, may follow.

It is desirable, therefore, to be aware of all possible causes of failures, the failure modes and the failure mechanisms. If one is armed with this knowledge, careful and systematic examination of the failed transformer and the environment will lead one to the root cause of the failure. It will also be possible to carry out reliability reviews of the designs and to carry out stage inspections during manufacture, in order to ensure that there are neither seed-defects nor gross defects in the transformers as built. This is the only way to attain the full potential of total reliabilty of transformers.

The causes of transformer failures, generally in order of their probability, can be classified as follows:

(a) Seed-defect or defect in an internal connection or a terminal. (Discussed briefly in Sections 4.4 and 4.7 in this chapter, and again in detail in Chapters 5 to 12)

(b) Failure of interturn insulation in the main winding due to:
- relative movement between turns (Section 4.6)
- local overheating due to defect in connector or terminal (Sections 4.3, 4.4 and 4.7)
- overheating due to overload (Section 4.2)
- local overheating due to obstruction in oil circulation (Section 4.7)
- local overheating due to sludge accumulation (Section 4.7)
- mechanical damage during manufacture (Section 4.5)
- local overheating due to proximity of steel and eddy currents due to stray magnetic fluxes (Section 4.7)
- inadequate balancing between and excessive heating of some of paralleled conductors
- sharp edges on copper conductor (defect in raw material) (Section 4.5)
- moisture penetration between turns (Section 4.5)
- tracking across wooden cleats (Section 4.4)

(c) Main insulation failure between the winding and the transformer tank due to:
- moisture entry into oil (Section 4.8)
- reduced clearance between flexible lead and tank (Section 4.8)
- tracking across wooden cleats (Section 4.4)

It may be noted that some of the failures which seem to be of the type (b) may be actually due to a defect of type (a). The insulation may have been damaged in such cases by overheating caused by bad connections. This distinction is very important because all insulating materials are very vulnerable to damage by overheating, and if the damage is due to overheating, there is no way to cure the problem except by preventing the original cause of overheating.

There are, of course, the other cases of transformer failures which are truly classified under group (b) in Section 4.3, which are due to failure of insulation for reasons other than local overheating caused by bad connections. These will be discussed in Section 4.5.

4.4 FAILURES DUE TO DEFECTS IN INTERNAL CONNECTIONS AND TERMINALS

Bad connections usually manifest themselves by the increase in contact resistance and the consequent overheating of the conductors. The heat developed in the joint between conductors is proportional to the product of the square of the current and the contact resistance. The conductor size is generally selected in such a way as to ensure that the temperature rise of the conductor is within safe limits, but due to economic considerations, the margin between the safe limit and the operating temperature obviously cannot be made very large.

There is a critical temperature level, which is well below the permissible temperatures for insulating materials, at which a vicious circle of increasing temperature and increasing contact power loss is established. Such a vicious circle culminates sooner or later in catastrophic failure of the transformer. There are no practical means of detecting such dangerous spots during service and the only possible way of preventing them from occurring is by taking certain precautions in the design, manufacture and installation of the transformers.

It is possible to detect and prevent overloading, to protect transformers against external surge voltages, to guard against failure of cooling equipment by monitoring the oil temperature, to monitor the condition of the transformer oil; but it is not practicable to detect local overheating at defective internal connectors and terminals. When such defects occur, sooner or later, failure of the transformer is certain.

Some of the usual locations where bad connections lead to transformer failures are enumerated below:
- Terminals where the external cables are connected to the transformer.
- The points on the terminals where the internal connectors are bolted to the terminals.

- The points at which the internal connectors are joined to the windings.
- The points at which the winding wires or conductors are joined to internal busbars or terminal boards or tapchangers.
- The points at which internal connectors or leads are gripped in cleats or pass through insulated partitions.
- The points at which contact is made between the moving contact and the fixed contact of the tapchanger.

In a 3-phase transformer, the number of vulnerable points of the types listed above is likely to be in the range 15 to 45 or even more. The number depends on design details and the type of internal tapchanger that may be provided. Each of these vulnerable points is a possible source of failure if any of the defects or seed-defects described in Chapters 5 to 12 are present in the equipment. As almost all these points are inside the transformer tank, they are not accessible to inspection or monitoring. If defects or seed-defects are present, there would be gradual degradation which would remain undetected until a catastrophic failure occurs; and even then, the real cause may not be revealed due to the extensive damage to the equipment which is likely to occur.

It was mentioned earlier, that the degradation which takes place in connectors and terminals inside the transformer tank would remain undetected until failure occurs. This statement needs to be qualified a little. It is possible, as a result of recent developments in condition monitoring systems, to detect such points by carrying out periodic dissolved gas analysis (DGA) on oil samples drawn from the transformer. However, this method is neither fool-proof nor completely dependable. If the degradation is very rapid, failure may occur even before the next DGA is due. Therefore, while, it should be carried out periodically to detect possible defects which may occur in spite of care being taken at the design, manufacturing, and installation stages, it should not be considered as a substitute for such care.

The tapchanger, usually of the off-load type, which is provided inside the transformer tank needs special mention. The foregoing discussion in this chapter and in Chapters 5 to 12 are very relevant; but there is an additional point which has to be checked. This is the actual tap connection between the fixed and moving contacts. The contact is usually spring loaded and the contact force is very important. It must not be allowed to fall below the specified limit as a result of wear or poor quality of springs. Type tests on the transformer must include measurement of temperature rise at 100 percent overload at the tapchanger contacts and terminals after a mechanical endurance test on the tapchanger. This temperature rise must not be more than 40°C.

A trivial type of defect in laminated wood cleats used for separating leads can be the root cause of a major failure. Due to the presence of a minute air gap between the layers of laminated wood cleats, tracking may occur either between two leads from a winding or between one lead and the earthed fixing bolt (see Fig. 4.2).

Failures of this type are basically due to improper impregnation of the laminated cleat and the entry of moisture into airgaps. Failures of this type can be prevented by:
- Baking new cleats at 100°C for 24 hours and storing them while hot in dry, transformer oil.
- Testing new cleats in oil for one minute at three times the anticipated maximum voltage between the conductors.
- Ensuring that transformer cores lifted for maintenance are kept out of oil for as short a time as possible.

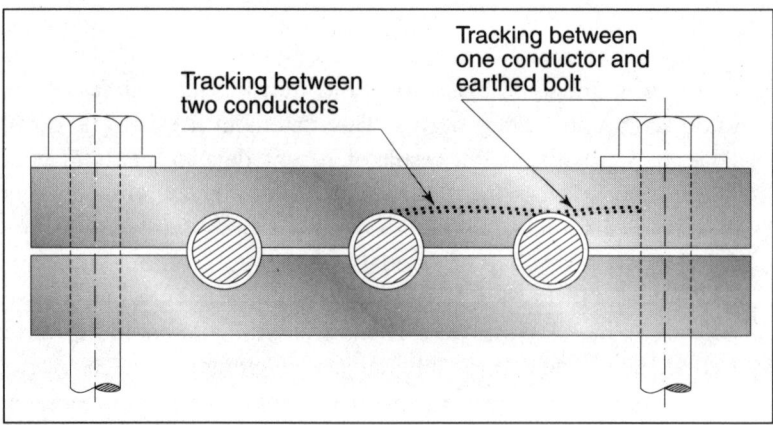

Fig. 4.2 | Failure due to tracking between laminated layers

4.5 FAILURES DUE TO INTERTURN SHORTS IN WINDINGS

There are several possible defects or seed-defects which can lead to interturn short circuits; but before discussing these in greater detail, the effects of such shorts may be examined first. Interturn shorts lead to very high circulating currents in the shorted turns. Due to the very high transformation ratio from full primary turns to a few turns, the increase in the primary current is likely to be very small despite the high current circulating in the shorted turns. Therefore, overcurrent relays in the primary circuit are unlikely to detect such faults. The fault would then persist unchecked. The excessive heat developed continuously would cause the oil around the shorted turn to vaporise. The sudden increase in volume may lead to an explosion.

Transformer windings are usually made of paper insulated rectangular conductors wound flat in the shape of pancakes. These are placed over one another around the magnetic core, but separated by insulating spacers. The insulation between two adjacent turns of the same pancake consists of four layers of paper impregnated with transformer oil. The interturn voltage, i.e., the maximum operating voltage between adjacent turns, is less than one twentieth of the breakdown voltage of new insulation. Therefore, under normal operating conditions, there is no possibility of failure of this interturn insulation.

The paper insulation of transformer winding conductors is, as can be imagined, mechanically delicate. It can be easily damaged by constant abrasion. It is also vulnerable to embrittlement and cracking if the operating temperature of the conductor goes beyond the permissible limit of 140°C at any point. The insulation can also absorb moisture and lose its insulating properties.

Figure 4.3 shows a typical transformer coil stack. Failure of interturn insulation is due either to damaged paper insulation or to dropping out of loose spacers. Such defects may occur due to one of the following reasons. The preventive action to be taken has also been indicated:

- If the bare copper conductors used in the manufacture of the paper insulated conductors have sharp edges on the corners, the insulation may get shorted during service under the effects of vibration, thermal expansion and contraction, movements caused by electromagnetic forces and even the static assembly force between the coils. Conductor specifications usually indicate the minimum radius at corners and surface finish of the conductors. These points have to be checked during stage inspection of the raw materials.
- Mechanical damage during manufacture of the conductor or during the winding and assembly of the transformer may reduce the strength of the paper insulation. The paper insulation is delicate while the coils are heavy. They have to be handled very

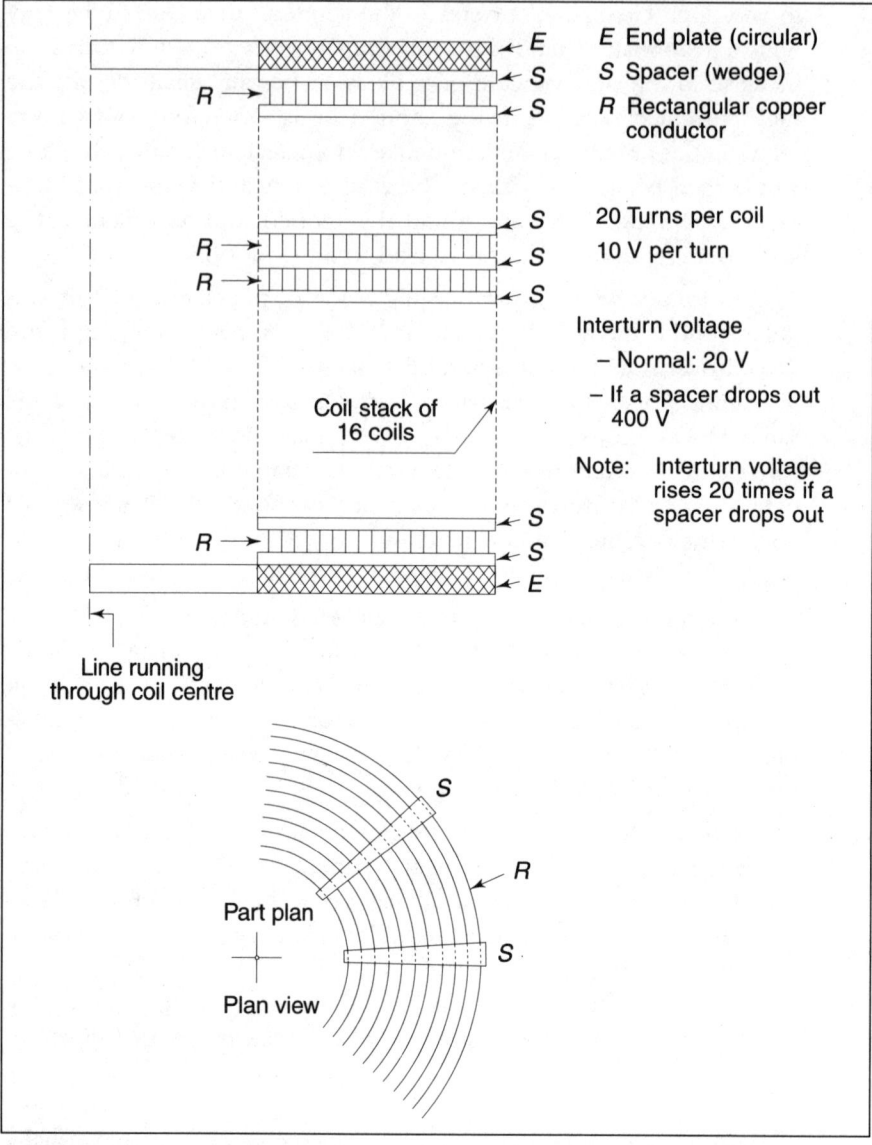

Fig. 4.3 In this sketch, only a few coils (one at each end and two near the middle) are shown. Similarly, only a few turns in each coil are shown. The effect, of coil spacers dropping out, on interturn voltage is also shown.

carefully during manufacture. If the transformers are obtained from reputed manufacturers, the likelihood of this defect is small. Nevertheless, the purchaser of the transformer must insist on the following tests and inspections:

- Type and routine checks and tests on the copper conductor and the conductor insulation.
- Double frequency, double voltage routine test on all the transformers.
- Impulse voltage test on the prototype transformer.
- The most common cause of damaged interturn insulation is relative movement between adjacent turns. This failure mechanism is discussed in detail in Section 4.6.
- Short circuits between turns may be caused by the presence of moisture in the paper insulation. During drying out operations of the transformer sufficient time must be allowed between completion of the drying out and final testing of the oil to ensure that any moisture which may have been trapped in the coils is given time to diffuse out into the bulk of the oil.
- Interturn insulation may be damaged also by thermal damage due to excessive oil temperature. This failure mechanism is discussed in Section 4.7.

4.5.1 Interturn Shorts Due to Relative Movement of Turns

It was mentioned in Section 4.5 that interturn shorts caused by relative movements between turns is a very common cause of transformer failures. These movements are caused by:

(a) alternating electromagnetic forces between the adjacent turns. These forces and movements are particularly high during momentary high peak currents caused by short circuits and faults in the 'downstream' installation or to high in-rush currents when switching ON.

(b) vibration in the case of transformers mounted on locomotives and motor coaches.

(c) thermal expansion and contraction due to the constant changes in temperature.

It is not possible to eliminate these three factors; but it is certainly possible to prevent any relative motion between the conductors. Special design features have to be provided for this purpose.

Failures of interturn insulation due to mechanical damage caused by relative movement of conductors is usually due to either non-provision of the special design features needed for this purpose, or incorrect adjustment or maintenance of the same.

The paper insulation on the conductors is easily damaged by abrasion, but it is very strong and resistant to failure when held under direct compression. Therefore, the method adopted to prevent failures of transformers due to interturn shorts caused by relative movement of the turns is to hold the coils together under very high compressive pressures. The pressure is so high that the coils are held immobilised and relative movement is not possible. Figure 4.4 shows the arrangement in a schematic form. This figure illustrates the principle. Actual details may vary from one manufacturer to another, but the principle is the same.

It will be seen from Fig. 4.4 that the coils are held under compression by a number of screws and springs. The dimensions and characteristics of these screws and springs are so selected as to ensure the required compressive force.

Relative movement between turns is also possible due to shrinkage of interturn spacers and lack of compensating springs. Improper or inadequate seasoning of the coils during manufacture is often the root cause of the failures.

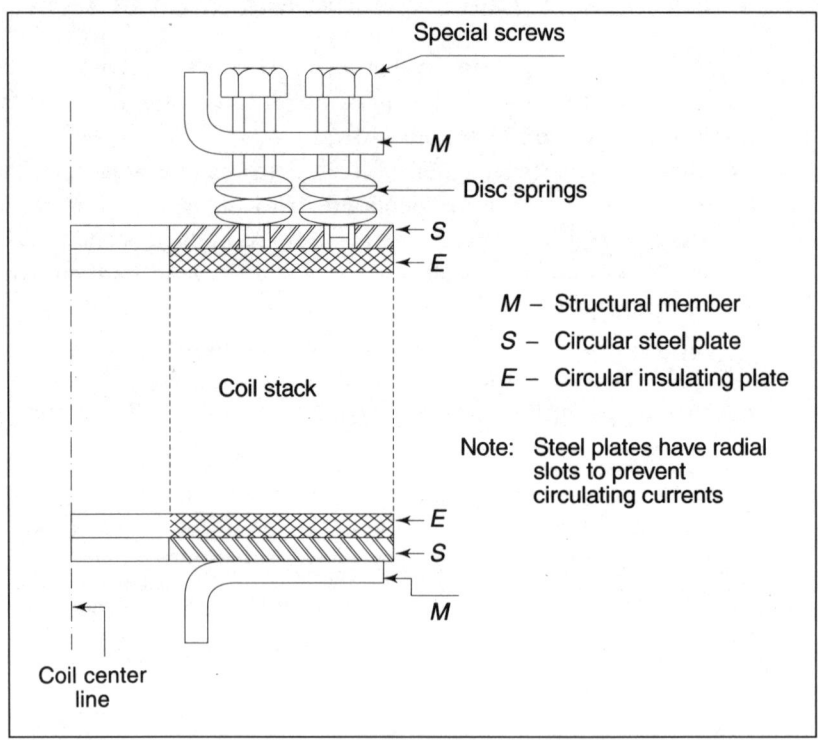

Fig. 4.4 Clamping of transformer coils to prevent interturn movement

Transformer Failures

The purpose of providing springs is to permit the thermal expansion of the copper conductors when they get heated up during service. If the springs are not provided, the copper will be subjected to very high compressive forces which are sufficient to produce a permanent set in the conductor thickness. This will cause the coils to become loose when cold.

In order to minimise the shrinkage of the paper insulation and the laminated spacers between the coils after assembly of the transformer, the following precautions have to be taken during the manufacture of the transformer coils.

- The laminated paper spacers are precompressed under high pressure so as to minimise further shrinkage during service. This has to be verified by an ageing test under heat and pressure as part of the acceptance tests. The compression set in this test must be within the specified limits.
- After assembling the coil-stacks, they are subjected to a curing process consisting of heating, pressing, cooling, pressing cycles in special fixtures until the coil stack height stabilises.
- The transformer assembly must include a suitable arrangement comprising a pressure plate with a number of screw/spring supports, to hold the coil stacks under high pressure at all times.
- Inspite of the three precautions listed above being taken, a small further reduction in coil height may take place after some years of service. During the first overhaul of the transformer after five to seven years, the coil pressure screws should be tightened until the springs are compressed to the designed lengths.
- In many transformers currently in use, the above precautions may not have been taken. As a result, the coil stacks and coil spacers are likely to have become loose. Some of the spacers may have even fallen out. Figure 4.3 shows how interturn voltage may increase twenty- fold if a spacer drops out. All such transformers are bound to fail.

4.5.2 Interturn Shorts Due to Thermal Damage to Insulation

Thermal damage to insulation due to overheating caused by overloading is a theoretical possibility to be considered only when there is definite evidence of overloading. While checking this point, it must be remembered that transformers have considerable capacity for short time overloading. The permissible overloading limits may be checked from the transformer specifications. If these are not exceeded by significant margins, and if the transformer has not already been in troublefree service for more than 20 years, overloading as a possible cause of failure should be the last point to be considered after eliminating all the other possible causes of failure.

Thermal damage to insulation due to overheating caused by failures of cooling equipment is also another possibility to be considered. The oil temperature measured after the transformer has been on full load for a few hours is a good indicator of the proper functioning of the cooling equipment. This should be compared with the prototype test results of the transformer. Whenever cooling equipment is provided, temperature sensitive trip relays are also provided and these may be expected to operate in the event of failure of cooling equipment, thereby preventing transformer failure.

Thermal damage to insulation may also be caused by local overheating due to poor circulation of oil as a result of some design defect. This possibility may be considered specially if the point of failure is the same in two or more transformers.

Thermal damage to insulation may also occur if sludge from the oil accumulates at the bottom and covers up some part of the winding. This would interfere with the free flow or convection of the oil and thus cause local overheating of the coil.

Thermal damage to interturn insulation due to overheating of connectors and terminals is actually a failure mode which is common, but often missed during investigations. Copper is a good conductor of heat. If any of the connections or terminals get overheated due to their own internal defects, the excess heat may get carried as far as the winding.

The paper insulation on the conductors is far more vulnerable to damage by heat than any other insulating material used in the transformer. It may be recalled that the conductor itself will get heated up due to its own resistance and its normal operating temperature is likely to be near the limiting temperature. A little extra heat conducted from a defective connection or terminal or a little less heat dissipation due to sludge accumulation, may easily cause irreparable damage to the paper insulation.

Therefore, whenever investigating a transformer failure, the position of the fault centre in relation to terminals and connectors should be a specially checked. There may be little evidence left as a result of the damage on the failed transformer. In such cases, the design details of a similar undamaged transformer may be checked with reference to the explanations given in Chapters 5 to 12.

Measurement of winding temperature during type and routine tests by the resistance method gives the average temperature of the coils between terminals. Reliability and durability of a transformer depend not on the average temperature, but on the hot spot temperature. Even in a well designed transformer, the difference between the hot spot temperature and the average temperature is likely to be of the order of 10°C. Allowance is usually made for this varition. However, defects in the detailed design can result in much higher differences. Some of these possible defects are:
- Narrow oil passages between some coils.
- Presence of 'hydraulic short circuit paths' for the flow of oil causing starvation of oil flow to certain sections of the coils.
- Obstructions to oil flow due to badly shaped or badly placed separators between coils.
- Presence of heavy steel structural components close to some part of the coil may lead to local overheating due to stray fluxes and eddy currents in the steel parts.
- Overheating in a badly designed or badly manufactured joint or terminal could lead to overheating and transmission of heat to the coil insulation.

There is no simple test possible to locate such hot spots. The only way to do so is through a very careful examination of the design drawings and the prototype. The oil flow paths should be marked and the estimated percentage of the total oil flow in each oil passage should be determined by a consideration of the relative resistance to flow of parallel paths. This has to be done during a reliability review of the transformer design before commencing manufacture and also if coil failures occur at the same location repeatedly in different transformers of the same make and design.

4.6 FAILURE OF THE INSULATION BETWEEN WINDING AND THE TANK

The main insulation between the transformer windings and the tank is provided largely by the mass of the transformer oil and oil impregnated paper-board laminates. Depending upon the operating voltages of the windings, adequate clearances have to be provided between the windings and the earthed tank. Since this insulation gets tested during routine hipot tests and since it can be easily verified visually, defects in this regard are rare.

Failure of the insulation can occur mainly as a result of ingress of moisture into the transformer oil. This ingress could be due to improper or defective maintenance or repair of silica-gel breathers. It could also occur due to cracks or leaks in the upper sections of the transformer tank, pipework and conservator. Such defects often go undetected, while similar defects in the lower or oil filled sections become obvious due to leakage of oil.

Transformer failures may also be caused by voltage surges due to lightning or switching; but this is only the proximate cause. The root cause is the absence of adequate arrangements in the design of the transformer and the protective gear.

If the lengths of flexible leads to terminals are excessive, they may swing close to earthed parts and this could lead to earth faults.

4.7 OTHER FAILURE MODES

The more common failure modes have been discussed in the foregoing paragraphs. Some of the other failure modes which are relatively less frequent are enumerated and discussed briefly in the following paragraphs.

Failures of transformers can occur due to gross design defects such as improper selection of rating, impedance (in case of parallel operation), or over-voltage capacity, although such cases are rare.

4.7.1 Failure of Core Bolt Insulation

Transformer laminations are usually held together under pressure by a number of steel bolts and nuts. In order to minimize eddy currents around the bolt shanks, heads and nuts, and to prevent circulating currents through two or more bolts, insulating sleeves are provided around the bolt shanks, and insulating washers are provided under the bolt heads and nuts (see Fig. 4.5).

If the insulating washers are not made of pre-compressed material, they shrink under the effect of oil, heat, pressure and ageing. This causes the bolts to lose their tension and the laminations to lose their compression. Noise and vibration follow. Mechanical damage combined with some original defect may lead to bolt insulation failures and local overheating. If adjacent coils are affected, they will, in due course, develop interturn shorts and, finally, transformer failure.

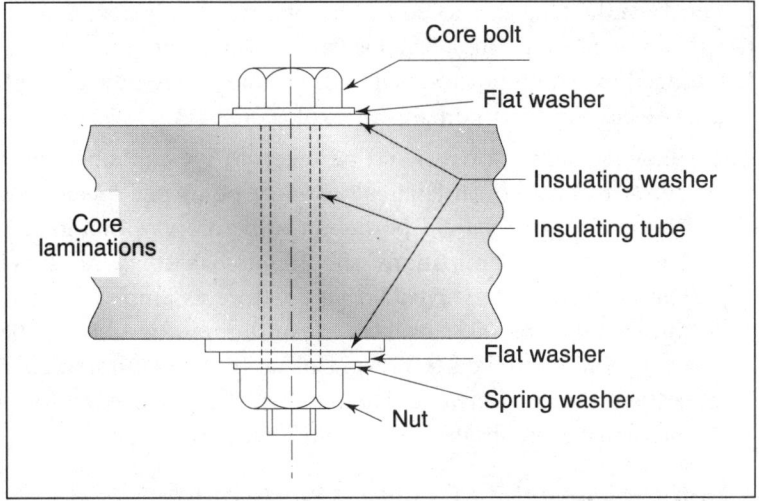

Fig. 4.5 Clamping of transformer laminations

The mechanism of failure described above may be a very slow process, but once it gets started, it is sure to end in transformer failure sooner or later. If the insulation on two adjacent bolts fails, the circulating currents will be so high that failure and probably a fire may start within minutes.

The following precautions during manufacture and maintenance should be self-evident from the above analysis.
- The washers under the bolt heads and nuts should be made of pre-compressed insulating laminates.
- The nuts should be tightened accurately using a torque wrench both initially and periodically during overhauls to the torque values specified by the manufacturer.
- The nuts should be adequately locked to prevent loosening in service.
- The bolt insulation should be checked initially and during overhauls with a 1000-V megger.

4.7.2 Failure of Insulation Between Laminations

The insulation between laminations is usually a very thin film of insulating varnish. If the varnish is not of the required quality or if it has not been applied and cured properly during manufacture, the insulation may fail in service. Similarly, if the laminations have burrs on their edges, the insulating films may get punctured. Either way there would be local overheating due to eddy currents. When this heat is transmitted to adjacent coils, interturn insulation will deteriorate and finally fail altogether. The ultimate result is the same: short circuit, overheating, gas formation and perhaps an explosion.

The following precautions should be taken to prevent such failures:
- The varnish used on laminations must be as per specifications.
- The laminations should be deburred before varnishing.
- The varnished laminations should be baked at the specified temperature for the specified time before assembly.
- The laminations must be kept clean during assembly. Particular care should be taken to prevent metal filings or turnings/borings getting lodged between laminations. They are sure to cause interlamination shorts and eventual failures.

4.7.3 Failures Caused by General Overheating Due to Overvoltage

The no-load current of a transformer is normally less than 10 percent of the full load current. This current serves mainly for the magnetization of the core and depends, therefore, on the magnetization characteristic of the laminated core. For economizing on the size and cost of the transformer, the normal operating point on the magnetization characteristic is close to the saturation point (see Fig. 4.6).

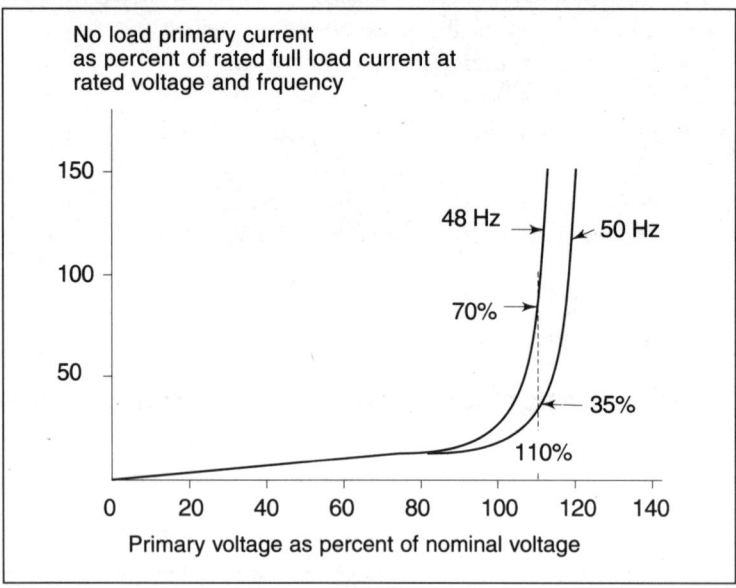

Fig. 4.6 Effect of frequency and voltage on no-load primary current

If the primary voltage rises beyond the maximum specified primary voltage, the no-load current rises by a much higher percentage. Thus, a 20 percent rise in the primary voltage above the maximum specified primary voltage may increase the no-load current to a level well above the full load current and the transformer may be severely overloaded even when the secondary or load current is well within safe limits. A similar effect is produced by a fall in supply frequency below the lowest specified frequency. Obviously, a rise in voltage coinciding with a fall in frequency is very dangerous.

Such a defective condition is shown up clearly by the ammeters indicating the primary and secondary currents. The secondary current, determined only by the load may be within limits; but the primary current which depends on the load as also the magnetising current would be considerably in excess of the safe limits. If overcurrent relays in the primary circuit or transformer temperature operated relays are provided, tripping of the circuit breaker will save the transformer. Otherwise, general overheating of the transformer will reduce its durability and in extreme cases, failure of transformer is also likely.

Overheating of the transformer may also occur if the supply wave form is flat topped and not sinusoidal. This is unlikely to be the sole cause of failure, but it may be a contributory cause when other defects such as high voltage, low frequency, cooling equipment failures, and overloading are also present.

Shell type three-phase transformers are vulnerable to problems created by third harmonics in the current generated by saturation of the magnetic core when the transformer windings are connected in star/star. Depending upon the earthing or isolation of the neutral points, either excessive current or excessive voltage of the third harmonic may be the result. This could also be a contributory factor in transformer failures.

Whenever a transformer is switched on, there is a high inrush curent. As the duration of such an inrush current is very small, its thermal effects on the windings are usually negligible. However, the electromagnetic forces produced by these short duration inrush currents can cause damage specially if the coils are not held firmly under pressure. High primary voltages or low supply frequencies lead to high flux densities in the magnetic circuit and these result in higher inrush currents. Frequent switching of transformers in the vicinity of power stations can lead to transformer failures due to movement of coils. The only way to prevent such failures is to ensure that:

- the transformer design caters to the maximum likely voltage and minimum likely frequency.
- the mechanical design of the winding and core is such as to keep the windings under adequate pressure at all times.

4.8 INVESTIGATION OF TRANSFORMER FAILURES

A very important part of any drive or effort to achieve zero failure performance of existing transformers with a poor record of reliability, is the investigation of each and every case of transformer failure that may occur.

While restoration of service must receive the first priority, the investigation of failure must be completed within a few days thereafter. Since failure has occurred, it is certain that there is a defect or seed-defect in one or more of the following factors:
- the specification
- the design
- the manufacture
- the installation
- the operation
- the maintenance

It is not sufficient to identify one or more of the above six factors. The exact defect or seed defect must be identified and it must then be checked and eliminated from the other transformers in the system. For such checks, priority should be given to transformers of the same make and type. Changes must also be effected in the documentation (such as specifications, drawings, test schedules, maintenance manuals, and training modules) relating to the component or equipment concerned, so that future procuremnent, operation and maintenance is done correctly.

As the methods of investigation are common to all equipment, these are discussed in detail in Chapter 19.

4.9 PROCUREMENT OF NEW TRANSFORMERS

The usual transformer specifications which are stipulated and referred to during the purchase and inspection of transformers do not cover any of the points mentioned in this book. These have to be verified by a reliability review of the detailed design of all the internal and external connectors, terminals and cleats to make sure that none of the seed-defects described in this book are present. Such a review may be made first at the stage of design drawings, and again at the stage of the inspection of the prototype. If necessary, an enabling clause for such a review may be incorporated in the specification.

The prototype tests stipulated in the specifications must include the following

- measurement of temperature at all junctions or joints on and between connectors and terminals, and
- mechanical endurance tests on tapchangers if any are provided.

4.10 DO'S AND DON'TS FOR INVESTIGATING TRANSFORMER FAILURES

If transformer failure rate is more than one percent per year, assume that there is need for a thorough review of specification, design, manufacture and maintenance practices.

Do not assume that overloading is the cause of failure, unless oil temperature at the end of the peak load period has been found to be consistently higher than 120°C.

Investigate every case of transformer failure, by lifting the core and carefully examining the insides to determine the root cause of failure.

If there is any burning or overheating at any internal, connnector or terminal, or if there is any coil failure near a connector or terminal, examine carefully for defects or deficiencies of the types described in Chapters 5 to 12.

Follow the procedures suggested in Chapter 19 regarding investigations of fires and failures.

4.11 DO'S AND DON'TS FOR PREVENTING TRANSFORMER FAILURES

- Ensure that raw materials used in the manufacture of the transformers comply with the revelant industry standards.
- Ensure that precompressed spacers or separators are used and that the coil stacks are subjected to curing cycles during manufacture until coil height is stabilized.
- Ensure that all the specicified tests in the standards for transformers are carried out on the new transformer. In particular, check the no-load current at the maximum specified operating voltage suitably corrected for the minimum specified frequency

(If the frequency at the time of the test is higher than this specified minimum).
- Ensure that coil pressure screws are provided in the design and that these are fully tightened during overhauls.
- Ensure that the transformer oil is maintained in accordance with industry standards for liquid-type (or liquid-filled) transformers. In particular, see that sludge is not allowed to accumulate to coil level.
- Ensure that all connectors and terminals have the desirable design features referred to in Chapters 5 to 12.

Chapter 5

Failures of Electrical Connectors and Terminals

In this chapter, we will discuss:

- Applications and types of electrical connectors.
- Significance of connector failures.
- Effects and criticality of connector failures.
- Failure modes of connectors.
- Causes of connector failures.
- Role of contact force and temperature.
- Measures for preventing connector failures.

A chart showing seven different failure modes of electrical connectors and terminals, and their 29 possible causes is given in this chapter while their details are discussed in subsequent chapters.

Although the use of aluminium conductors and connectors is not recommended, the precautions to be taken when their use is unavoidable are listed in this chapter.

5.1 INTRODUCTION

While discussing electrical fires and failures in Chapter 3 about electrical equipment and installations, and in Chapter 4 about transformers. it was stated that many such cases are caused by seed-defects in electrical connectors and terminals. Yet, these seed-defects are not viewed with the necessary seriousness partly because most of them continue to exist in service without causing any problem and partly because all signs of the few seed-defects which do develop into fires, are obliterated by the effects of the fire.

The electrical connector is the simplest type of component used in electrical engineering systems and one would expect it to be totally reliable. Yet, the fact is that failures of electrical connectors do occur from time to time and most of the failures are due to apparently trivial and avoidable defects in design, manufacture or installation of this simple component.

It is necessary, therefore, to enumerate and to describe in detail the different types of seed-defects in electrical connectors and terminals; but before doing so, it would be useful to describe the common and special features of different types of connectors and terminals.

5.2 TYPICAL ELECTRICAL CONNECTORS

The majority of electrical installations consist of various electrical machines or equipment connected together by wires and cables. A typical distribution substation consists of the following equipment:
- Incoming transmission line
- Lightning arrestors
- Current transformers
- Potential transformers
- Main power transformer
- Primary switchboard
- Secondary switchboard
- Battery
- DC distribution panel
- Battery charger
- Metering, control and protection equipment
- Outgoing cables

These pieces of equipment, which may be 10 to 20 in number, would be connected together by cables or wires. Most of these would be insulated, but some may be bare copper wires or busbars. The cables would generally be made of stranded copper wires and may vary in cross-section from 2 mm^2 to 300 mm^2 or even larger.

Figure 5.1 shows a schematic power-circuit diagram of a small distribution substation with two incoming and four outgoing cables and two transformers. It shows the main power connections, but not the control connections between the various major equipment. The

Failures of Electrical Connectors and Terminals 87

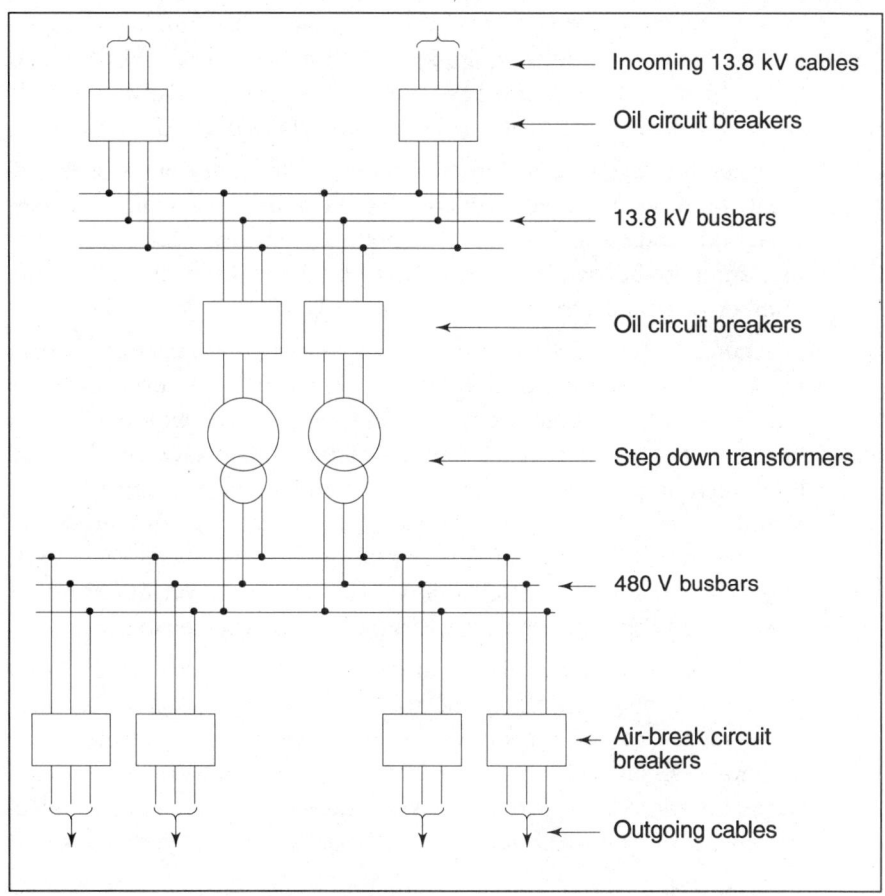

Fig. 5.1 Schematic power circuit diagram of a small 13.8 kV/480 V substation

total number of electrical connectors is likely to be of the order of 200, and the number of bolted, crimped and spring loaded contacts, in the power circuit only, is likely to be more than 600 even in this small substation. The number of connectors in the control circuits is likely to be double this figure, i.e., about 1200.

In order to provide proper electrical contact and to facilitate installation and maintenance, each piece of wire or cable has to be provided with terminal fittings or sockets made of tubular copper. These terminal fittings are fixed with threaded fasteners on the terminal boards of the equipment.

It would be seen that even a small substation has hundreds of connectors between various equipment. Each connector would have two bolted joints and two crimped joints. These joints involve pressure contacts which can, if badly made, become over-heated during service.

Inside the casings or cabinets of these electrical machines or equipment, various components are connected together in a similar manner.

In an average substation, the number of such electrical connections may run into thousands. In large units like thermal power stations, the number of connectors would run into hundreds of thousands.

Figures 5.2 and 5.3 show two typical electrical connectors used in industrial electrical installations. This chapter and the seven chapters which follow are about electrical connectors similar to those shown here. In the vast variety of types, designs, and makes of electrical equipment in general use, there would be differences in regard to the shape and size, but basically, the electrical connectors considered here would consist of an insulated wire fitted with sockets or lugs at each end. The discussions which follow will be mainly on questions of features or design parameters which are relevant to reliability.

This chapter and the seven chapters which follow, are mainly about the failures of electrical connectors used in high power electrical installations of all types such as electric and diesel-electric locomotives, substations, power stations, heavy industrial drives, etc. Although there is a wide variety of sizes, shapes, lengths, socket designs, insulating materials etc. the basic features of design as also

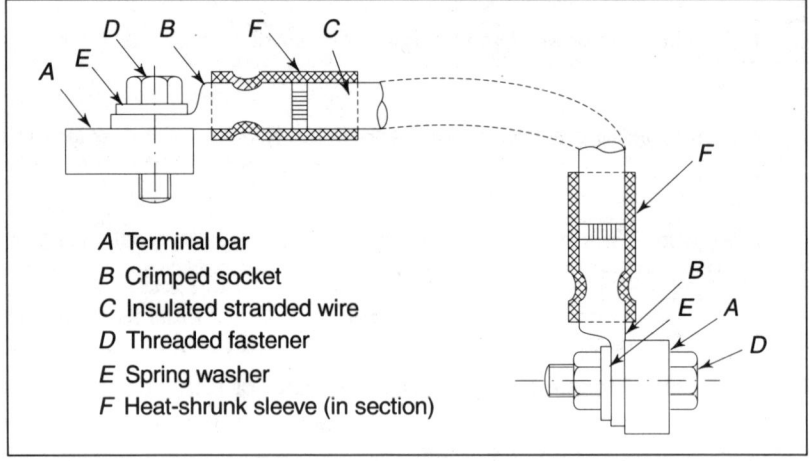

Fig. 5.2 Typical electric connector (with bolted connections)

Failures of Electrical Connectors and Terminals

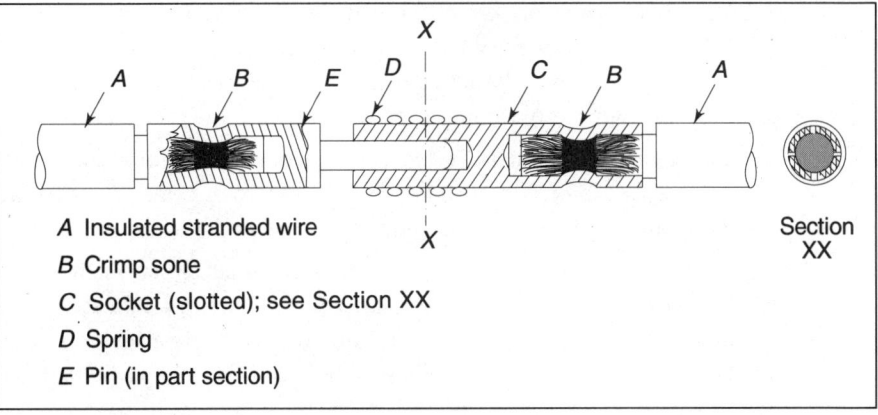

A Insulated stranded wire
B Crimp sone
C Socket (slotted); see Section XX
D Spring
E Pin (in part section)

Fig. 5.3 Typical electric connector (with plug/socket connections)

the failure modes and mechanisms are common to all types of connectors used to these diverse applications. Seven failure modes and 29 causes of failures are enumerated in Table 5.1 in Sec 5.11 of this chapter and discussed in Chapters 6 to 12. They would cover more than 95 percent of all failures of electrical connectors which usually occur in practice.

Electrical connectors used in the computer and electronic industry are not within the scope of this book. Due to the low voltages and currents used in these equipment, fires are rare, but many failures which take place are due to causes which are very similar to those in medium and heavy electrical equipment. In case of hardware problems in computers or similar electronic equipment, many engineers check the connectors for loose contacts before going on to check for component defects.

The basic principles which apply to questions of reliability of the connectors in computers and electronic equipment are the same as those discussed in this book. The dimensions of components are generally smaller, the voltages and currents are lower and some of the materials used are different; but the modes and mechanisms of failures are identical to those encountered in the failures of electrical connectors used in medium or high power electrical installations.

5.3 APPARENT TRIVIALITY OF CONNECTORS AND THEIR DEFECTS

The worst disasters are often due to minor defects in small components. As most of such failures result in extensive damage, it is difficult to distinguish cause from effect. The cause of failure is often attributed to failures of more sophisticated equipment in the vicinity, usually because it is generally believed that complex equipment are more likely to fail than simple components.

Sometimes, attention is diverted to more complex equipment deliberately because no one in authority would like to attribute a major fire already in the public eye to the failure of a bolted joint!

The electrical connector is indeed in a peculiar situation. It is neglected not only by the designers and manufacturers, but also by the maintenance staff; and as if to justify this neglect, it is also ignored at the stage of investigation of failures. The result of all this is that the root causes of many electrical equipment failures are not exposed and totally avoidable failures continue to take place.

To illustrate the point that connector failures are not really as trivial as they might seem to be, we may consider a practical example. The most expensive component in a high power rectifier cubicle is the diode. In any high power high voltage rectifier cubicle, there are many diodes, connected in series-parallel, but the failure of even one diode gets detected by the protective system, and the faulty section of the circuit is disconnected from the source of power within a fraction of a second, before any damage can occur to the other components. The rectifier can be put back into service after replacing the defective diode.

On the other hand, the failure of a bolted connection of relatively negligible cost can lead to a costly fire in the rectifier cubicle and perhaps even in the locomotive or substation which houses the rectifier cubicle. In this case repairs are very expensive; and it may take weeks or even months before the equipment is restored. This happens mainly because there is no practical or economical device now available for detecting overheating in bolted or crimped connections and for automatically disconnecting the defective section of the circuit from the source of power.

Most electrical equipment run unattended. If even one of the thousands of bolted or crimped connections develops a defect, it may steadily get worse without anyone becoming aware of it. Eventually, it may become so hot as to ignite the insulating material around it, and this may start a fire which may spread and destroy the whole installation. This is not a mere hypothetical case. Many such fires occur regularly. The majority of electrical fires which are usually attributed to short circuits not only in the popular press but also in some technical

Failures of Electrical Connectors and Terminals 91

reports are actually caused by electrical connector failures of the type described in this chapter and in Chapters 6 to 12.

The apparently trivial problem of electrical connector failures is aggravated into a major issue by several factors, necessitating a detailed discussion. These factors are:
- There may be hundreds or even thousands of electrical connectors of various types in an electrical installation.
- All these connections are generally made manually and the quality of the joints depends on the care with which the joints are made.
- The usual or common types of defects which may be introduced during design, manufacture or installation are rarely such as to cause immediate failure. They are actually seed-defects which may take months or even years before becoming active and causing failures.
- Seed-defects are doubly dangerous because many visible seed-defects continue to remain in service without causing any problem, thereby creating a false sense of security amongst the staff. Eventually, it is always one of these seed-defects which, after lying dormant for long periods, suddenly becomes active and causes failures of the equipment.
- There is no simple routine method available for checking the quality of each bolted or crimped connection either initially or periodically during service.

There is a possible method for detecting bad connections. The method involves the use of an infrared video camera with a VDT monitor. Any connector or joint which is operating at a high temperature becomes clearly visible on the screen. It is then possible to attend to such a defect in its nascent stage, thereby averting a failure or fire. However, the equipment required for this purpose is expensive and not easily available.

It is also possible to use colour changing stickers or crayons and contact thermometers to detect excessive temperatures of terminals, but special precautions have to be taken if these are used.
   (i) The equipment must be made dead before applying crayons/stickers.
   (ii) The temperature limits should be carefully selected after measuring the normal operating temperatures. These are often well below the specified limits.

(iii) Contact thermometers with insulated probes must not be used at voltages higher than the insulation level of the probes.

(iv) Reliability and durability of the stickers/crayons should be verified.

Many engineers find that it is quicker and more effective to tighten the fasteners periodically, instead of making periodical temperature measurements. During scheduled inspections, it is also desirable to keep a sharp look-out for signs of overheating such as discolouration, softening and charring of insulating material that is close to terminals and connectors.

It is therefore necessary to take special steps to see that electrical connectors are designed, manufactured and installed in such a way as to preclude all possibility of failure. The measures or precautions to be taken for this purpose are simple and practical, but perhaps it is their very simplicity which makes them liable to be neglected.

5.4 EFFECTS OF FAILURES OF ELECTRICAL CONNECTORS

Generally, electrical connector failures in control circuits are perceived and reported correctly. There is little damage and very little arcing because voltages and currents are low. It is usually possible to determine the complete failure mechanism and the root cause of the failure. The effect of control circuit connector failures is that it becomes necessary to shut down the system until the faulty components are located and replaced.

When electrical connector failures occur in power circuits, there is usually extensive damage due to arcing, burning and fire. Often, only lumps of molten and congealed metal and ashes of insulating material are found. It is difficult to decide where the first arcing or flashover took place. Such failures are rare, but the costs of damages in each case are very high.

Since short circuits are usually detected and isolated within a fraction of a second by the protective system of relays and circuit breakers, fires due to straight short circuits are rare and then they are due to defects in the protective equipment. On the other hand, overheating or burning of defective electrical connectors cannot be detected until it is too late and a fire has already started. At that stage, a short circuit may occur and the protective system may cut off the power supply, but then it may be too late and it would not be the cause of the fire.

5.5 CRITICALITY OF FAILURES OF ELECTRICAL CONNECTORS

Electrical connectors consist of varying lengths of insulated wires or cables with lugs or sockets at each end. These are simple components but the number of failures of electrical equipment and machines caused by their failures are large enough and their effects serious enough to make this a critical issue.

Failures of electrical connectors must be viewed and investigated seriously for several reasons such as:
- The consequential damages, including the loss of production which follows and the cost of repairs which have to be carried out, are often very great. The total costs may be several thousand times the cost of a connector. Sometimes, connector failures, can even be the cause of electrical fires, in which case the costs can be enormous.
- In some installations, there are thousands of electrical connectors and their overall failure rate is often very high. Although a very small percentage of these failures may cause major fires, almost every other case will lead to failure of the main equipment, necessitating its shutdown until repairs are carried out.
- There is so much damage around the seat of failure, that it is usually not possible to determine the true cause of failure by examination of what remains of the failed component.
- Defects in electrical connectors are totally preventable and it is entirely within the scope of the currently available skills and materials to achieve zero failure performance from electrical connectors.

If electrical connector failures continue to occur from time to time, it is only because of the lack of awareness not only amongst the artisan staff but also amongst the supervisors and some engineers, of certain basic principles of electrical contacts, metal creep, and metal fatigue.

5.6 PREVENTIVE MEASURES

The basic principles in so far as they apply to the causes of failures of electrical connectors are discussed here, as well as measures for achieving zero failure performance from electrical connectors.

It will be seen that some of the failure modes and failure mechanisms are not self-evident and need to be taught through special training courses. While technically qualified people like supervisors and engineers would be able to grasp the points about design merely by going through these explanations, it would be necessary to run special courses for the electricians. For those who are involved in the inspection and installation of electrical connectors of different types, it will be necessary to impart training regarding the common types of defects in materials and in workmanship which lead to failures in service.

With proper training, even illiterate artisans can grasp the basic principles involved here. When artisans understand how certain defects can slowly grow and develop into failures and why they have to take certain precautions, they will do their work correctly, willingly and with enthusiasm. Training courses are important because all electrical connectors are installed manually. There are no routine tests that can be done on each connector and a great deal depends on the workmanship. Bad work is usually due to lack of awareness of the critical or important details of the work amongst the staff. It does not take any more time or effort to do the work correctly than to make the usual errors which lead to failures in service.

Inspection of installed electrical connectors by foremen or supervisors prior to energizing circuits is one important way of ensuring the safety and quality of installations. However, since 100 percent inspection is rarely practical in the field, training electricians and technicians to terminate conductors correctly is the most practical way to achieve zero failure performance.

5.7 TRUE OR ROOT CAUSES OF CONNECTOR FAILURES

Many major failures of electrical equipment and machines which are now being ascribed to more complex problems are actually due to the failures of electrical connectors. This is because the true or root causes of failures are not immediately apparent when many components and equipment are damaged. It is often difficult to distinguish between the cause and the effect.

When several components are damaged, there is a natural tendency to assume that the failure started in the more complex component. For instance, if an electromagnetic contactor burns out, the defect is at once assumed to be in the contactor itself, whereas the defect may have been in the electrical connector socket bolted on the terminal of the contactor. Since many failures result in the burning of equipment and connected cables, many investigations proceed on the wrong track. In general, the apparent simplicity of the electrical connectors tends to prevent adequate attention being paid to them not only during design and manufacture, but also during investigation of failures.

5.8 THE IMPORTANCE OF CONTACT FORCE

The basic requirement for good electrical contact between two components is the mechanical force between them. A large number of failures of electrical equipment is due to the ignorance or neglect of this simple fact.

Electrical connectors can be divided into two classes on the basis of the manner in which the contact force is developed:
- connectors in which the contact force is developed by threaded fasteners.
- connectors in which the contact force is developed by springs, or by elastic deformation as in crimped sockets.

There are other ways in which electrical connectors can be classified, for instance, on the basis of control or power, low voltage or high voltage; but for our purpose of the study of reliability, the classification given above is the important one.

5.9 CONTACT FORCE DEVELOPED BY SCREWS OR NUTS/BOLTS

Although solid copper wires or bars are sometimes used for connectors, the most common type of connector consists of a stranded copper wire or cable fitted with crimped sockets at each end, as shown in Fig. 5.2.

The length of the wire between the crimped sockets may vary from a few centimeters to several hundred meters, and the cross section of the wire may vary from a few to several hundred square millimeters. In general, the crimped sockets at each end of the connector are fastened to terminal boards, busbars or other sockets with the aid of screws or nuts and bolts. The wires may sometimes be bare, but, generally, they are insulated.

The force necessary for obtaining good electrical contact is developed by tightening the screw or bolt/nut provided for this purpose.

(See Chapter 15 for a detailed explanation on the relationship between the mechanical force between metal components and the quality of the electrical contact.) This is perhaps the most important but least understood aspect, which is relevant not only for electrical connectors, but also for many other electrical equipment such as relays, contactors and switchgear.

5.10 CONTACT FORCE DEVELOPED BY SPRINGS

When it is necessary, to disconnect and reconnect the connector from time to time on account of operational or maintenance needs, plugs and sockets are used at the ends of the connectors. The force required for obtaining good electrical contact is provided by springs. A common example of this design is the 3-pin plug and socket used for supplying electric power to domestic electrical appliances. Similar but larger fittings are used in heavy industry and railways mainly for control cables and occasionally for power cables. One example of a multicore cable connector with plug/socket terminations is shown in Fig. 5.3.

There are many different designs of multicore plugs and sockets, but they all have a few common features. The 'pins' and the 'sockets' are connected to stranded wires, which are usually insulated, through crimped joints. The sockets have spring loaded contact segments. The free bore of the socket is smaller than the diameter of the plug pin by a carefully controlled amount. This is done by specifying certain tolerances on the dimensions of the mating components. When the plug and socket are pressed together, the springs around the contact segments have to bend or deflect and this provides the force which is so essential for obtaining good electrical contact.

In a crimped socket, good electrical contact is ensured by the forces developed between the wire and the socket due to the elastic component of their deformation during the process of crimping.

5.11 MODES AND MECHANISMS OF ELECTRICAL CONNECTOR FAILURES

The technology involved in the design, manufacture and installation of electrical connectors is quite simple and there are only a few precautions which need to be taken to achieve zero failure performance from electrical connectors. These are discussed in detail

Failures of Electrical Connectors and Terminals 97

in the subsequent chapters, but the more common failure modes and their causes are listed in Table 5.1.

Table 5.1
Common Failure Modes and Their Causes

Failure Mode		Causes of Failure	Chapter
Overheating and/or burning of the crimped joint between the cable and the socket	(i)	Inadequate crimp	6
	(ii)	Incorrect dimensions of socket or crimping dies.	
	(iii)	Lack of tinning or oxidation.	
	(iv)	Inadequate penetration of wire into socket.	
	(v)	Incorrect location of crimp on socket	
Burning of the insulation around a spring loaded plug and socket	(i)	Overheating and/or weakened spring	7
	(ii)	Incorrect spring characteristics.	
	(iii)	Plug and socket dimensions incorrect.	
	(iv)	Defective crimp.	
Fracture of the crimped socket	(i)	Sharp bend or stress raiser	8
	(ii)	Excessive crimp.	
	(iii)	Vibratory flexing of the wire.	
	(iv)	Minor deficiency in material of socket.	
Fractures of wire strands	(i)	Excessive crimp.	9
	(ii)	Vibratory flexing of the wire.	
	(iii)	Sharp corner on barrel of socket.	
	(iv)	Absence of shrunk sleeve over wire and socket.	
Failure of insulation of the wire	(i)	Wire bearing hard on structural member.	10
	(ii)	Inadequate support for wire.	
Failures of terminal boards	(i)	Inadequate tightening of threaded fasteners.	11
	(ii)	Shrinkage of boards.	

	(iii)	Current passing through steel bolts.	
	(iv)	Insufficient tracking clearance.	
	(v)	Metal Creep.	
Failures of welded, brazed and soldered joints	(i)	Inadequate cleaning.	12
	(ii)	Improper materials.	
	(iii)	Improper process parameters.	
	(iv)	Hydrogen embrittlement.	
	(v)	Metal creep.	

It is necessary to refer here to a type of failure which is sometimes taken to be a connector failure, whereas the damage to the connector is actually only the effect of a certain type of failure of the terminal board caused by a defect in its design or manufacture. In such a case the terminal gets overheated. As the crimped socket at the end of the connector is in good thermal contact with the terminal, the former also gets overheated. In some cases, the overheating may be sufficient to burn the insulation on the wire and this may lead to a short circuit or fire. Generally however, only the insulation around the terminal gets burnt and the connector insulation is not affected.

The six failure modes for connectors listed above and the failure mode for terminal boards would cover more than 95% of all connector failures and all these are entirely avoidable (see Chapters 6–12 for the precautions or measures to be taken for preventing such failures.

The most reliable type of electrical connection between two conductors is the brazed or welded joint; but this type may also fail in service due to defects in design or process (also see Chapter 12).

5.12 FAILURES DUE TO MAJOR DESIGN DEFECTS

In enumerating the possible modes and mechanisms of failures, it has been assumed that there is no defect or deficiency in the general design of the connector on considerations such as temperature rise and insulation strength. For instance, it is assumed that the cross-section of the cable and the dimensions of the socket are not too small for the maximum current to be carried by the connector, or that the insulation thickness is not too small for the voltage to be withstood. These assumptions are generally quite valid because:

- designers usually follow certain proven empirical formulae given in handbooks or manufacturers' catalogues.
- Any design errors get detected during prototype tests which are usually made to verify the design.

If the failure rates are too high and there are doubts about the general design of the electrical connectors, it is desirable to check the design criteria.

A few of the possible major design defects which can cause failures are:
- Inadequate cross section of the wire, for the current to be carried, resulting in a general overheating of the connector.
- Inadequate cross section of the fastener terminal bar, or busbar for the current to be carried, resulting in overheating of the connector terminals.
- Inadequate insulation strength of the cable in comparison with the operating voltages, causing insulation failures.
- Improper material of the connector, busbar or terminal bar resulting in overheating of the connector.

Defects or deficiencies in regard to the above four factors are generally rare and have been enumerated only for the sake of completeness of the list of causes of failures. In actual practice failures are due to other less obvious seed-defects which, will be discussed in detail in the following chapters.

5.13 FAILURES DUE TO OVERHEATING

It is possible that in some cases, the initial temperature rise may be high due to major design defects such as those listed in Sec. 5.12. In general, standard design practices regarding current densities ensure that the temperature rise limits on connectors are not exceeded, but under special conditions, it is possible that at some locations they may develop higher temperatures due to high temperature in the vicinity of the connectors. Some examples of such special conditions are:
- The ventilation around the component may be restricted.
- There may be some heat generating components situated close to the component under consideration.
- Eddy currents in nearby steel components may cause heating in the vicinity.

It is desirable, therefore, to carry out temperature rise tests on all or at least some typical bolted or crimped joints at the end of the continuous rating test of the equipment. Such tests are a must if there are unexplained failures of terminals or connectors. The measurements should be made under worst-case operational and environmental conditions.

Design criteria are beyond the scope of this book but one very important criterion needs to be emphasised here. The permissible temperatures in degree Celsius at the contact surfaces are given in Table 5.2.

Table 5.2

Permissible temperatures at contact surfaces

	Bare copper	Tin-plated copper	Silver-plated copper
Bolted joints	90	105	115
Contacts	75	90	105
Crimped sockets	70	85	

The differences between bare copper, tinned copper and silver-plated copper are on account of the different rates of oxidation of different types of surfaces and different resistivities of oxidised surfaces. It is desirable to design the cross sections of the conductors in such a way as to get operating temperatures which are at least 10°C lower than the limits given above in order to obtain zero failure performance under worst case conditions.

It may be noted that these permissible conductor temperature rises on considerations of bolted joint reliability are much lower than those permissible for the windings on the basis of insulation reliability. The latter can be in the range of 130–180°C for class B and class H systems, respectively. This factor is important where a bolted or crimped terminal or joint is directly on or close to a winding which is designed to operate at temperatures which are well above the limits of 85–90°C for crimped or bolted joints. Heat may be conducted to the joint, causing it to heat up to a temperature at which the joint degradation processes (discussed in Chapters 6 and 7) may become active.

The reasons for allowing higher operating temperatures for tinned and silver-plated contact surfaces may be stated briefly here. All metals except noble metals like gold or platinum get oxidized in air even at room temperatures. A thin film of oxide develops on the surface of all such metals; but the rate at which this film is formed, the thickness of the film and the conductivity of the film vary from metal to metal. They also depend on the temperature. The magnitude and stability of the contact resistance depends on all these factors. Research and practical experience has shown that for ensuring reliable operation, the temperature limits given in Table 5.2 must not be exceeded.

Although aluminium terminals and conductors are used in many installations, permissible operating temperatures have not been given here because this metal is just not capable of giving reliable pressure contacts. There are far too many technical problems. Some of these are as follows:

- Aluminium is subject to metal creep at room temperature. At higher temperatures, the creep rate rises to unacceptable levels.
- Aluminium develops a hard and high-resistance oxide film within seconds even at room temperature.
- Bimetallic action between copper and aluminium creates additional heating effects.
- At high temperatures, as in an electric arc, aluminium is combustible.

It is best, therefore, to avoid the use of aluminium conductors. However, if its use becomes unavoidable, the following special precautions should be taken in the design, manufacture and maintenance of such conductors:

- As far as possible, use crimped sockets to terminate all conductors. Ensure that socket and crimping die design is proved by endurance tests (see Chapter 6).
- Use steel bolts and nuts, but ensure that there is direct contact between the conductors being connected together.
- Use plain steel washers and disc springs to compensate for metal creep and to maintain the contact force developed by the fasteners.
- Before assembly, clean the contacting surfaces with vaseline-impregnated steel wool. The purpose of the vaseline is to prevent or minimise re-oxidation of the contact surfaces after cleaning and during service.
- Assemble the bolted joints immediately after cleaning without removing the vaseline, which gets squeezed out by the contact force.
- Tighten the bolts or nuts with the specified torque initially, and retighten periodically. The interval between retightenings may be determined by trial and error initially. If there is no movement

of the nut at the specified torque, the interval may be increased for the next schedule.
- Check temperature rise periodically at rated current to make sure that it remains stable.

5.14 DO'S AND DON'TS FOR PREVENTING FAILURES OF ELECTRICAL CONNECTORS AND TERMINAL BOARDS

- Do not consider failures of electrical connectors and terminal boards as trivial failures. They are often responsible for major failures and fires.
- As there are no practical methods available for the detection of defective connectors and terminals in service and for automatic disconnection of supply, do take special care in their design, manufacture and installation.
- Since failures of electrical connectors and terminals can be totally prevented by taking a few simple precautions, do aim at zero failure performance.
- Since electrical connectors and terminals are installed manually, do arrange for the systematic training of the staff who are assigned this type of work.
- Ensure the intrinsic reliability of electrical connectors and terminals by using proven designs. Ensure reliability in the field by adequate quality control on materials and processes.
- For details of measures to be taken refer to the Do's and Don'ts at the ends of Chapters 6 to 12.
- If the use of aluminium conductors is unavoidable, do take the special precautions listed in Sec. 5.13.

5.15 CONCLUSION

Special attention needs to be paid to details regarding design, manufacture, installation and maintenance of equipment. Ignoring any of these does not mean that there would be immediate failure. Whether failure occurs or not, depends on many other factors such as the vibration levels, general stress levels in the design, intensity of thermal cycling, etc. Yet, if it desired to achieve zero failure performance, all these aspects must be kept in view.

Failures of Electrical Connectors and Terminals   103

The measures proposed in this chapter do not add to the cost of the equipment, but they improve the reliability by several orders of magnitude. When there are thousands of bolted or crimped connections in any installation, it is necessary to insist on certain minimum standards for design, manufacture, installation and maintenance of electrical connectors and terminals. If this is done, it is possible to achieve zero failure performance from electrical connectors and terminal boards.

Chapter 6

Overheating/Burning of Crimped Sockets

In this chapter, we will discuss:

- The first and the most common of the seven failure modes of electrical connectors and terminals, viz. overheating/burning of crimped sockets, which is the result of one or more of the following six defects or seed-defects:
  - inadequate crimp
  - incorrect dimensions of socket or of crimping die
  - lack of tinning or oxidation
  - inadequate penetration of the wire in the socket
  - incorrect location of the crimp on the socket
  - defective material of socket

- The crimped socket system of terminating cables, together with its advantages over the system in use earlier, viz. soldered sockets.

- The mechanisms or processes of failures which follow the presence of each of the six defects or seed-defects and the preventive measures.

6.1 INTRODUCTION

Overheating and burning of crimped sockets is the most common type of failure in electrical connectors, especially when heavy currents are involved. Since the socket is usually destroyed due to arcing which follows the failure of the socket, the real cause of the failure is often not determined.

Overheating/Burning of Crimped Sockets

If the current carried by the socket as well as the operating voltage is low, the small arc that is produced when the failure actually takes place may get extinguished by itself, and in such a case, there would be no fire. Of course there would be some interruption in the service or in the operation, the severity of which would depend on the exact position of the interruption in the circuit.

If such a failure occurs in a high voltage circuit, or if the current being interrupted is high, the resulting arc will be so intense that it may not get extinguished on its own. If there is an earthed component in the vicinity, the arc will divert to that component, which may cause the current to increase momentarily, thereby actuating the overcurrent protection system. The situation is now similar to that when an insulation failure is followed by a short circuit. The protective system will cause the main circuit breaker to open, thereby quenching the arc. If the total time lag between the starting of the arc and the opening of the circuit breaker is small, there will be no fire.

In the absence of a grounded component in the vicinity of the failed socket, there is a real possibility that the arc which is formed may continue to flare. As the current in the arc would be only the normal operating curent or even less, and as there would be no ground leakage current, there is no way by which the protective system could come into action. This is the most dangerous situation and the failure is certain to escalate into a fire.

This type of failure is responsible for many fires which take place in heavy electrical installations. Fortunately, these defects can be totally eliminated by taking a few simple precautions during the installation of the cables and sockets.

There is one other mechanism of failure in which there is no parting between the socket and the wire, no arcing, short circuit or ground fault. In this case, there is only overheating of the socket due to excessive internal contact resistance. Under normal conditions, i.e., when there is no defect, the internal resistance of the socket is such that the heat generated in it is within limits and the socket temperature is below 80°C. If there is some defect in the manner in which the crimping is done, the internal contact resistance may rise to five, ten or even more times the normal level. In such a case, the socket temperature would rise to several hundred degree Celsius. The insulation on the cable near the socket would also get heated and may even ignite at such temperatures. If this happens, a fire is certain to follow.

Actually, amongst the various methods of terminating cables, the crimped socket system is the most reliable and is therefore used almost universally in all electrical installations in the transport and manufacturing industries, and in electric power stations. Crimping of sockets onto wires can be done very quickly on the shop floor with the help of special crimping tools which are easily available.

Even when the failure rates of crimped sockets do not seem to be excessive when calculated on a per hundred socket basis, or when compared with the failure rates of other systems of terminating wires, they can be unacceptably high. Considering the very large number of sockets used in each installation and the fact that socket failures may lead to costly fires, it is necessary to aim at zero failure rate.

6.2 THE SOCKET SYSTEM OF TERMINATING CABLES

The wires or cables used for connecting together various components inside one or several equipment in an installation are usually stranded copper wires ranging between a few to a few hundred square millimetres in total cross-sectional area.

In order to ensure that all the strands of the wire are properly connected together at each end, it is necessary to provide a socket or lug at each end of the wire. The socket is usually made from a copper tube as shown in Fig. 6.1. The use of a socket to terminate a cable makes it possible to connect and disconnect it quickly, as and when required. This facilitates assembly, trouble-shooting and maintenance.

Sockets are generally made from annealed high-conductivity copper tubes. Part of the tube is flattened and drilled so that it can be bolted onto a terminal board or another socket. The insulation on the cable or wire is stripped and all strands are gathered together and inserted into the barrel of the socket.

There are two common methods of connecting the cable end to the barrel of the socket—tin soldering and crimping.

6.3 TIN SOLDERED SOCKETS OR LUGS

Tin soldered sockets are no longer in general use. However, in case it becomes necessary to use this system, the correct method of fitting such sockets is given below. The process also shows how much quicker it is to provide crimped sockets.

Overheating/Burning of Crimped Sockets 107

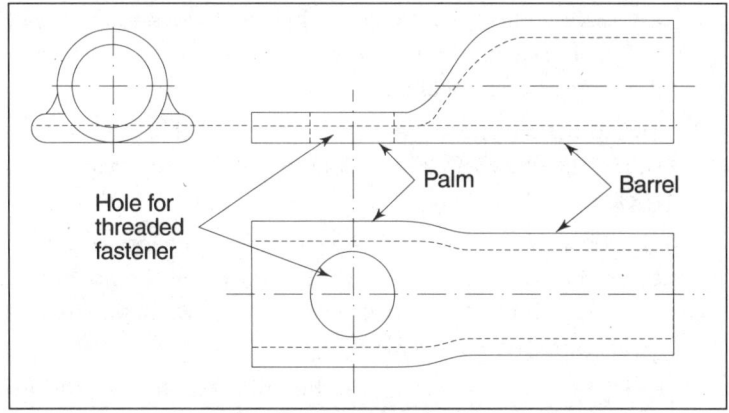

Fig. 6.1(a) Lug or socket for terminating stranded wire

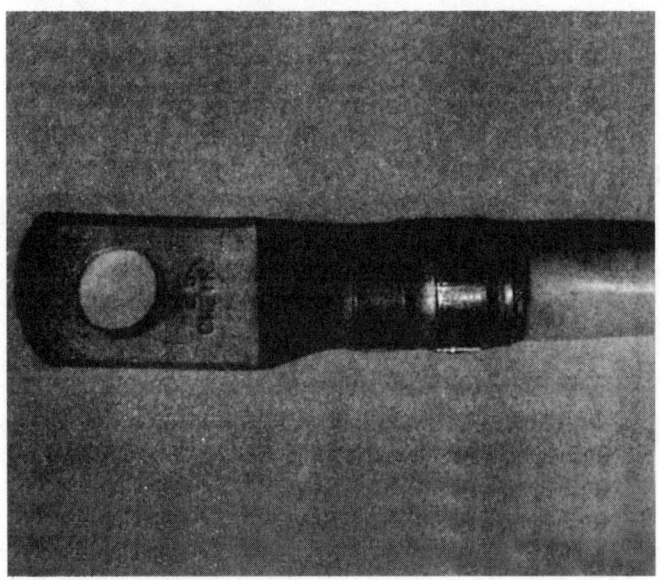

Fig. 6.1(b) Socket crimped on a cable

- Degrease the stripped end of the stranded cable if necessary, by dipping in unused perchloroethylene.
- Similarly, degrease the barrel of the socket in unused perchloroethylene. (This type of degreasing is not necessary when using the crimping process.)

- After allowing the perchloroethylene to evaporate, apply resin based, non-acidic flux to the inside of the socket barrel and to the strands of the cable.
- Tie up the stripped end of the stranded cable with a thin copper strand as shown in Fig. 6.2(a) to facilitate easy insertion in the socket.
- Tin the prepared end of the cable by dipping in molten tin at about 300°C. Pre-heat and tin the inside of the socket barrel by pouring molten tin into it with a ladle. Remove the excess tin by tapping the socket while hot.
- Hold the socket vertically as shown in Fig. 6.2(b), and insert the cable end into the barel while both are still very hot by the

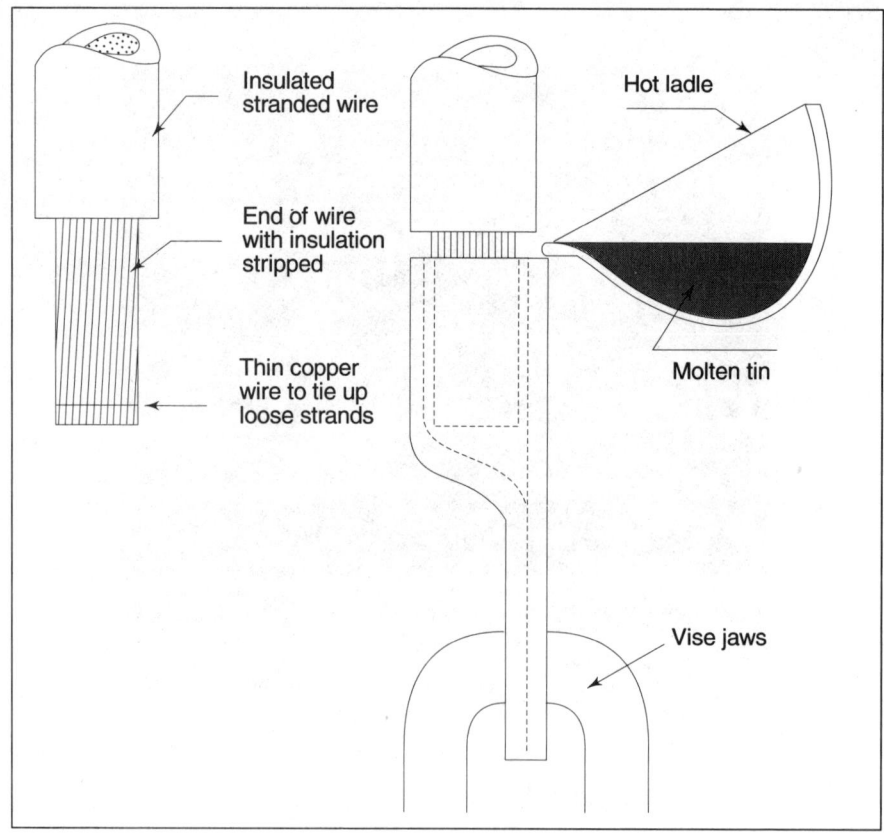

(a) Stripping of insulation from end of stranded wire

(b) Soldering of stranded wire in socket

Fig. 6.2

tinning process. Pour the molten tin slowly into the space between the wire and the socket barrel with the help of a ladle. Wipe off excess tin on the external surface of the socket while still molten.

- Hold the socket and cable immobile until the tin solidifies and the temperature drops to 100°C. A fan may be used to cool them rapidly.

Even when soldered with all due precautions as above, tin soldered sockets are intrinsically unreliable because tin is subject to metal creep at normal operating temperatures and its mechanical strength is poor. If due to any reason, there is any continuous or alternating force on the cable, the metal bond of the solder is likely to develop cracks. Such forces can appear either due to vibration or due to thermal expansion and contraction. The contact resistance will increase and this would then lead to a rise in temperature and eventual failure, although usually after many years of service.

6.3.2 Crimped Sockets or Lugs

In the case of crimped sockets, good electrical contact is produced by the internal mechanical forces due to the elastic deformation of the socket barrel and the strands of the cable. These forces are not affected by the passage of time because the creep temperature limit of about 135°C of copper is well above the operating temperature of about 80°C. Even when there are some external mechanical forces between the socket and the cable, the bond between them is not affected because the fatigue endurance limit of copper is not exceeded. Crimped sockets are therefore intrinsically more reliable than tin soldered sockets.

If a crimped socket Fig. 6.3(a) is sectioned through the centre of the crimp area, the appearance of the cut ends will be like that of a solid conductor [Fig. 6.3(b)]. The gaps between the wires and the socket disappear completely. If the cut section is polished and examined under a microscope, it would be seen that the wires have been deformed into close fitting polygonal shapes as shown in Fig. 6.3(c).

The crimped socket system requires the use of special tools. Small sized cables and sockets can be pressed together by hand operated tools, but larger sized cables require portable hydraulic presses. Figure 6.4 shows a hand operated crimping tool used for smaller cables, and Fig. 6.5 shows a hydraulic press-type tool for large cables.

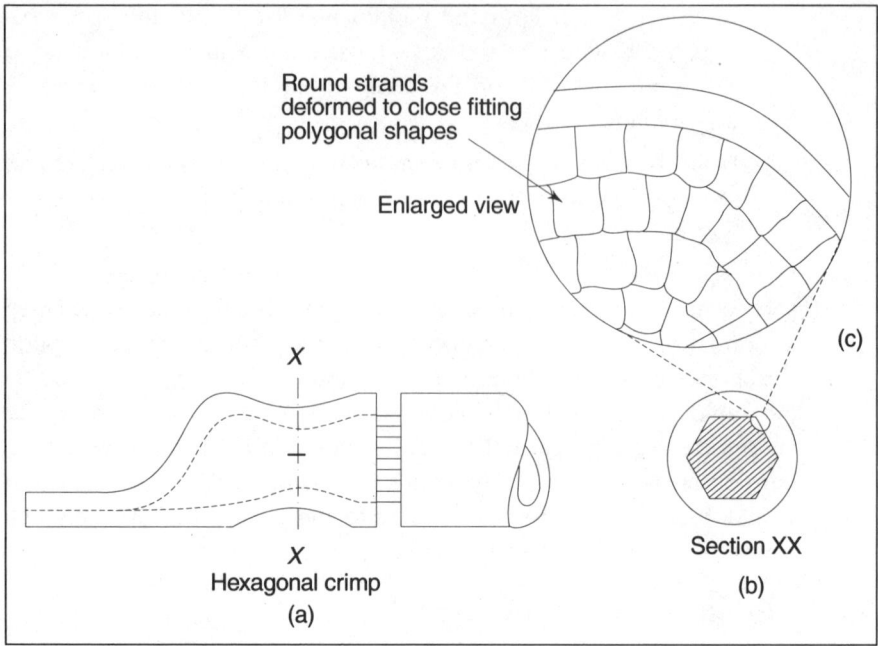

Fig. 6.3 Cross section through crimped socket

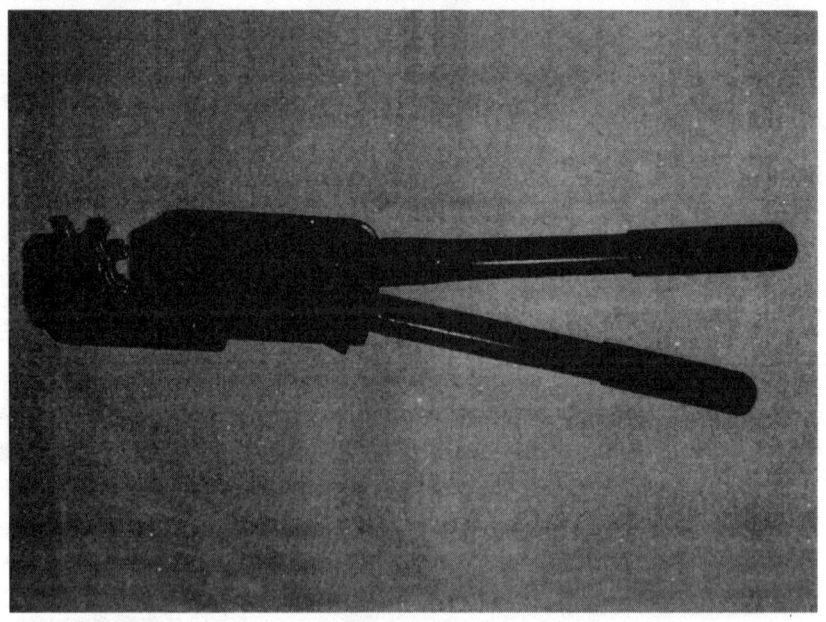

Fig. 6.4 Hand operated crimping tool for small cables

Overheating/Burning of Crimped Sockets   111

Fig. 6.5 Hydraulic press type crimping tool with hand pump

The crimping tools are so designed as to compress the copper of socket and of the wires by about 10–15 percent of their original area. This removes all air gaps, and at the same time, produces an elastic deformation of both the socket and the wires. Internal forces are developed between the wire strands and the socket, which ensure that the electrical contact resistance between the wires and the socket is kept at the minimum possible level. The role of contact force for producing good electrical contact has been discussed in Chapter 15.

The quality, reliability and durability of crimped joints depend on the following main factors:
- The shape and dimensions of the crimping tool dies.
- The shape and dimensions of the crimp socket.
- The quality of the material of the socket.
- The force developed by the crimping tool press.
- Complete penetration of wire in the socket.
- Correct centering of the crimp on the socket.

6.4 MODES AND MECHANISMS OF SOCKET FAILURES

If failures of crimped sockets are detected at an early stage, signs of overheating such as softening or charring of the wire insulation may be seen. However, since most terminal boards are not easily visible to the operating staff, it is more likely that the defect will reach the stage of a fire or short circuit before it is detected. The maintenance staff should therefore look out for such signs during periodical inspections of the terminal boards and connectors.

The normal contact resistance of a crimped socket is generally less than the resistance of a cable of length equal to the barrel length. The cross section of the barrel is about the same as the conductor and the temperature rise of the socket is very nearly the same as the conductor itself, viz. 30 – 40°C. The socket and conductor temperatures will generally be less than 75°C under full load conditions.

If there are impurities in the copper used for fabricating the sockets, the conductivity could be significantly lower and the temperature rise would then be correspondingly higher.

If the crimp is defective, i.e., if the contact force between the conductor strands and the barrel is less than the required level, the contact resistance will increase in inverse proportion. This relationship is discussed in greater detail in Chapter 15.

Overheating and/or burning of crimped sockets is usually the result of excessive contact resistance between the cable strands and the barrel of the socket due to a defect in the original crimp. The failure mechanism is as follows:

- If the material of the socket is defective or if the initial crimp is defective, the contact resistance between the strands and the barrel of the socket is excessive.
- The heat developed in the socket, being proportional to the contact resistance, is also higher than normal.
- The temperature rise of the socket, being proportional to the heat developed in the socket, is thus higher than the permissible level.
- When the temperature of the socket is higher than 85°C, the contacting surfaces of the strands and the socket get oxidized more rapidly.
- The oxidation of the contact surfaces increases the contact resistance further and this takes the process back to step (a) with increased intensity.

A vicious cycle is thus established, and eventually, the socket gets heated to a level at which the insulating material in contact with the wire may char and catch fire. In an extreme case, the joint may fail altogether, and an arc may be established. There would then be melting of copper and ignition of the insulating material in the vicinity. This vicious cycle of the five-stage process leading to socket failures is shown in Fig. 6.6.

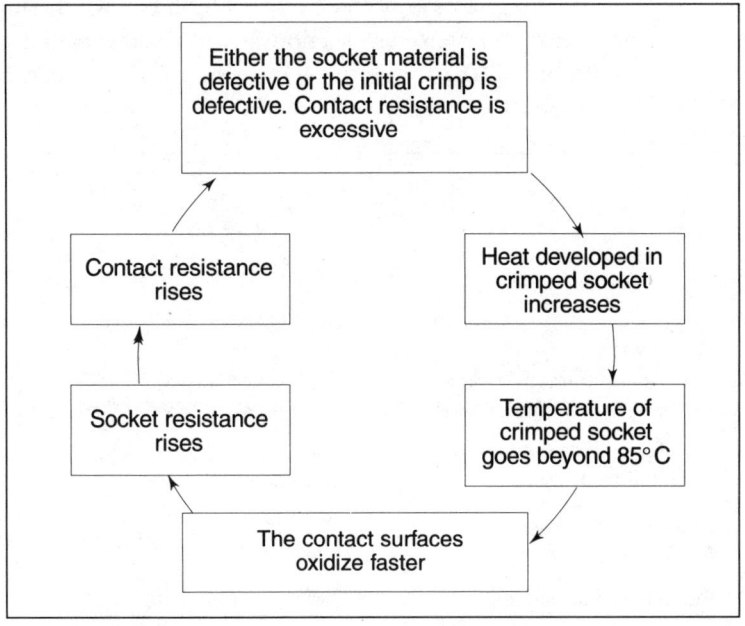

| Fig. 6.6 | The five stage cyclic process of socket failures |

When the socket temperature rises above the safe level of 85°C, the cycle of the five-stage process continues until failure occurs due to burning of the socket or ignition of the surrounding insulation. The only way to prevent such a failure is to ensure that the initial crimp is sufficiently strong, the contact resistance is low and the temperature is stable at well below 85°C. The design of the socket should be such that the contact resistance of the socket does not increase in service.

There are several possible ways, enumerated below, in which the initial crimped joint can develop high contact resistance. The proximate cause in every case is inadequate force between the cable strands and the barrel of the socket.
- The bore of the socket barrel and/or cable strands are not tinned and are badly oxidised even before crimping and use in service.
- The bore of the socket barrel is too large for the size of the wire.

- The thickness of the barrel being inadequate, the contact force exerted by the socket barrel on the wire strands is inadequate.
- The crimped cross section is excessive, and there is insufficient crimp due to use of incorrect or worn crimping tool or due to inadequate compression of the crimping tool. Fig. 6.7(a) shows a correctly crimped socket in section at the middle of the crimp zone, and Fig. 6.7(b) shows the section of a defective crimp due to the applied force being inadequate.

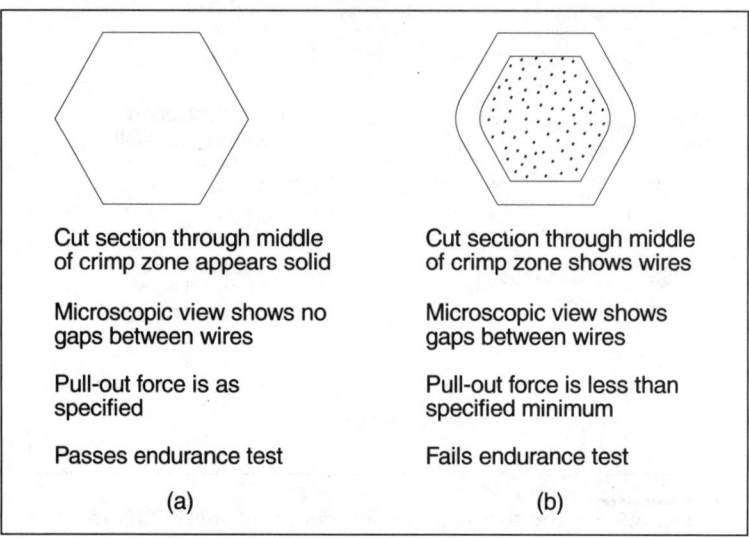

Fig. 6.7 (a) Cross section through a correctly crimped socket
(b) **Cross section through a defective crimped socket**

As an example, consider a stranded wire of cross section 50 mm^2, whose outer diameter is about 10 mm. If the bore of the socket is 10 mm and its outer diameter is 14 mm, the cross section of the barrel is

$$(14 \times 14 - 10 \times 10) \times 3.14/4 = 75 \text{ mm}^2$$

The total section of the wire and the socket is thus 50 + 75 = 125 mm^2.

The area of the outer diameter of the barrel of diameter 14 mm is

$$14 \times 14 \times 3.14/4 = 154 \text{ mm}^2$$

This has to be reduced by crimping to about 106 mm^2, because the minimum cross section of the socket at the center of its crimp zone should not be more than 85 percent of 125 (i.e., 106 mm²).

Overheating/Burning of Crimped Sockets

If, due to an error in the design of the crimping tool or due to insufficient force exerted by the crimping tool, this reduction is not obtained, the internal forces between the conductor strands and the barrel will not be sufficiently high to ensure a low contact resistance.

Even when the socket, the crimping tools and the crimping process are all correct, a defective joint can result if the size of the wire is not correct. A workmanship error which has often been detected is the cutting off of some of the strands to facilitate insertion of the strands into the socket. This not only reduces the conductor section in a critical area, but also makes the crimped joint defective due to inadequate compression of the copper.

It must be emphasized that the socket design, wire design and the crimping die design with regard to dimensions are important as a combination. Any change in even one of these, affects the quality of the crimp and the performance of the joint with regard to millivolt drop, pullout force and reliability.

6.5 · MEASURES FOR PREVENTING OVERHEATING/BURNING OF SOCKETS

The starting point for achieving total reliability, after the intrinsic reliability of any item is ensured by correct design, is the inspection at the purchasing stage of all raw material and components. In case of sockets, this means that special care has to be exercised with regard to the following:
- The copper for the sockets must have the specified conductivity, hardness, ultimate tensile strength, elongation and freedom from hydrogen embrittlement.
- the dimensions of the socket must be within the specified tolerances.
- The surfaces should be properly tinned.

The failure mechanisms mentioned above suggest the remedies to be applied or precautions that need to be taken to prevent such failures. These include:
- The dimensions of the socket, the type of crimping tool and the size of the wire should be matched in such a way that the minimum area of the crimp is about 85 percent of the total area of the socket barel and the cable strands. This means that the copper gets compressed by about 15 percent in the crimp zone, thereby developing adequate contact force between the cable strands

and the socket. As it is not easy to verify these figures, it is customary to make practical endurance tests as follows:
- Thermal cycling endurance tests, in which each cycle includes alternate heating and cooling, should be carried out to prove the design. A reference may be made to the relevant standard for crimped sockets to get the details of the endurance tests. A brief summary of tests given in BS 6360 is given below for understanding the principle of the test.
  (i) The resistance of a sample length of cable with two crimped sockets, as shown in Fig. 6.8, is measured. The resistance per unit length of the cable is also calculated by making measurements on a long piece of cable. The crimped joint resistance is then taken as half the difference between the resistance of the sample shown in Fig. 6.8, and the calculated resistance of the length of the cable between the sockets.
  (ii) The sample is then subjected to a temperature cycling test. In each cycle, a heavy current is passed through the sample until its temperature rises to 90°C, and is then cooled to 30°C, in a stream of air. This sequence is repeated 500 times. The resistance of the crimped joint is measured at ambient temperature and calculated as in para. (i) above after every 100 cycles.

The requirements of a good crimped joint are as follows:
  (i) The initial resistance of each joint should not be more than the resistance of a conductor of length equal to the barrel of the socket.
  (ii) The final resistance of the crimped joint after 300 temperature cycles should not be more than 150 percent of the initial resistance of the joint.

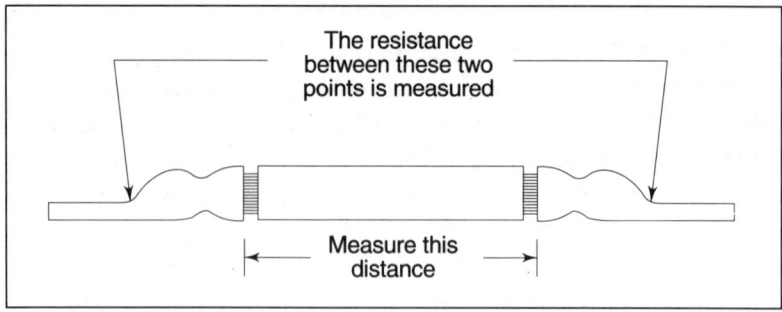

Fig. 6.8 Test sample for endurance test on crimped sockets

(iii) The resistance of the joint should not increase during the last 200 of the 500 temperature cycles.

(iv) Tensile tests should be carried out on the specimen to determine the loads at which one of the two sockets is pulled out. This pull out load should not be less than certain specified percentages of the breaking loads of the conductors. These specified percentages depend on the cross section of the conductor and the type of the conductor (i.e., solid, stranded etc.).

- Once the design of the socket and the crimping tool is proved by the above tests, there should be no changes in any of the design parameters such as dimensions and material of the wire and the socket, the dimensions of the crimping tool and the crimping force. Periodical checks should be made on the tools and on sample crimped joints.

The endurance tests have to be made only when selecting a suitable crimped socket and crimping tool design. When such a selection is made, the manufacturer of the crimped sockets and the crimping tools should be placed on the approved list, and the dimensions of the cable, the crimped socket and the crimping tools/dies should be frozen. Thereafter, it is necessary to carry out the endurance tests only if there is any change in the manufacturer or the design of the socket, crimping tool or cable.

Another check that should periodically be made by supervisors or by the stage inspection wing in production units, is the dimension of the crimped socket at its crimp zone. A suitable Go/Not-Go gauge may be made for this purpose. These checks are necessary because crimping tools can wear out, crimping presses can become defective and artisans can make mistakes, specially if they have not been properly trained.

While it is the duty of staff who carry out the inspection at purchasing stage to ensure that the sockets comply with the drawings and specifications, the shop floor staff who do the fitting can provide a second line of defence against defective supplies. They should be trained to report to the supervisors if they observe any unusual visible feature, any unusual clearance between the wire and the socket or anything unusual in the appearance of the crimp zone.

NECA/AA standard 104-1999, *Recommended Practice for Installing Aluminium Building Wire and Cable* (ANSI), provides guidance on termination of aluminium conductors. If the use of aluminium conductors

Chapter 7

Failures of Plug/Socket Connectors

In this chapter, we shall discuss:

- Applications where plug and socket connectors are generally used.

- Their advantages and special features.

- Their common constructional details.

- Significance of contact force and pull-out force between plugs and sockets.

- Seed-defects which develop into failures of plug and socket connectors.

- Their usual failure mechanisms.

- Practical measures for preventing their failures.

- As the key to reliable and safe operation of plug and socket connectors is in the provision of adequate contact force between the plugs and sockets, the methods used for carrying out acceptance tests for measuring this force will be discussed in detail.

7.1 INTRODUCTION

Plug and socket connectors are used widely in locomotive control circuits where the currents and voltages are low and frequent connection and disconnection is required. Low voltage multicore cable couplers are used between railway coaches for the transmission of power from generating coaches to the passenger coaches, mainly on account be trained to carry out pull out tests every day before commencing work. For this purpose, special steel blocks of the required weights may be made available. The worker has to make a sample crimped joint with a small piece of wire, and use it to lift the prescribed weight carefully. If there is any defect in either the socket or the crimping tool, the joint will slip out.

6.6 OVERHEATING DUE TO TERMINAL BOARD FAILURES

Sometimes, a crimped socket may get overheated and even burn due to a defect, not in the socket, but in the bolted connection between the socket and the terminal. Such a failure becomes more probable if the material of the terminal board is more resistant to heat than the cable insulation. This is a very common type of failure, often mistaken for a crimped socket failure.

As a connector socket is bolted on to a terminal board, it is not only in good electrical contact, but also in good thermal contact with the terminal. Hence, if the heat generated in a defective terminal is transmitted to the connector, under certain circumstances, the connector insulation may get damaged or the crimped socket may become overheated beyond 85°C and then fail due to its own further deterioration. Therefore, whenever a crimped socket fails, the possibility of the defect having originated in the terminal should also be checked up. (The causes of terminal failures are discussed in Chapter 10.)

6.7 WIRE WORKING OUT OF SOCKET

As mentioned earlier, overheating and burning of crimped sockets may be the end result of excessive bore of socket barrel, inadequate thickness of socket barrel, or of inadequate crimp. If the magnitude of these defects is beyond a certain limit, the failure is likely to be somewhat different in nature. The wire may work out of the socket.

There are two other possible defects which would lead to similar results. These are shown in Figs. 6.9(a) and 6.9(b). It may be noted that in small sockets, the positional errors as shown, have to be only of a magnitude of 1 or 2 mm to cause failures.

All the defects described above will have the same end result—the wire will work out of the barrel of the socket. Vibration and the initial tension in the wire will determine how long it would take for this to

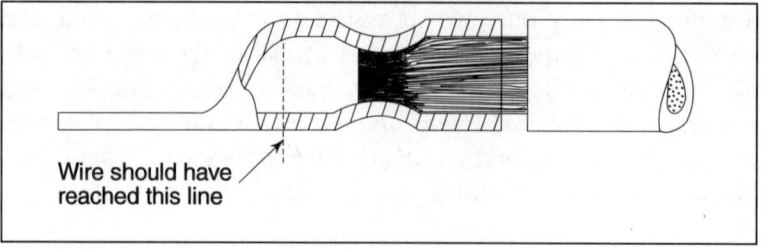

Fig. 6.9(a) Defective crimp where penetration of wire into socket is inadequate

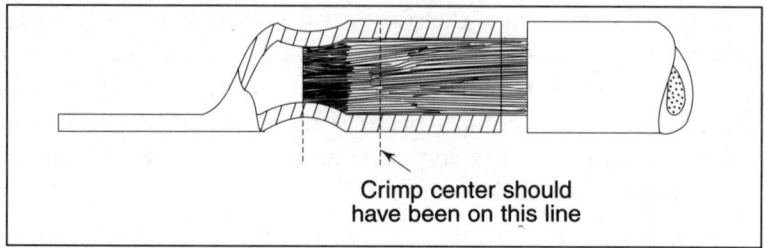

Fig. 6.9(b) Defective crimp where the location of the crimp is not in the center of the socket

occur. If the wire finally works out while electric current is passing through the joint, there would be arcing and possibly, a fire. Failures of this type in small cables used in control circuits do not result in fires. It is then easy to determine the root causes of the failures.

6.8 DO'S AND DON'TS FOR PREVENTING OVERHEATING AND BURNING OF SOCKETS

The steps to be taken to prevent overheating or burning of crimped sockets or working out of wires from the sockets can be summarized as follows in a list of do's and don'ts:
- As far as possible, use tinned wire strands and tinned crimping sockets. Tinning adds a margin of about 10°C to the limiting temperature of 85°C.
- Ensure that the barrel of the socket has adequate thickness and that the dimensions of the socket and the design of the crimping tool are proved by endurance tests on crimped joints.
- Ensure that the crimp depth and the crimping force are always adequate, i.e., not less than that used during the proving tests.

Overheating/Burning of Crimped Sockets

- The design of the terminal board should also be checked if the overheating and burning seem to have originated in the terminal board.
- Ensure that the wire penetrates the barrel fully and that the crimp is located in the centre of the barrel.
- Ensure that no change is effected in either the design of the socket, the design of the crimping tool or the size of the wire after any particular combination is proved by pull out and endurance tests.
- Ensure that the crimping tools and the crimping dies are verified periodically to ensure that there are no defects due to wear and tear.
- In order to prevent damage to the tools, ensure that they are stored with care when not in use.
- Ensure that the sockets are inspected carefully at the time of procurement. The dimensions, the quality of the material and the trial crimps should be checked carefully.
- Ensure that the staff who use crimping tools are trained fully in handling them.
- Ensure that the sockets are inspected carefully at the time of purchase with regard to the material and dimensions.

6.9 CONCLUSION

Overheating and burning of crimped sockets is probably the most common type of failure amongst the various types. It usually occurs due to incorrect choice of the crimping sockets and tools and poor maintenance of the crimping tools. They are more common in circuits in which the currents are high.

Failures due to overheating and burning of crimped sockets can be completely eliminated by taking a few simple precautions in the selection of the socket and the crimping tools, and in the actual crimping process. Sometimes, socket failures may occur due to defects in the terminal boards, but these are also capable of being totally eliminated, (discussed in Chapter 7).

cannot be avoided, follow NECA/AA-104, but for maximum reliability aluminium conductors should be avoided because this metal is subject to metal creep, bimetallic action with copper and rapid oxidation at operating temperatures. These three physical properties greatly reduce the intrinsic reliability of joints with aluminium conductors. Where the use of aluminium conductors is unavoidable for any reason, very special care has to be taken initially and also periodically during maintenance thereafter. The additional precautions to be taken with the use of aluminium cables and sockets are:

- Just before crimping a socket onto a cable (or bolting a socket onto a bar), apply joint compound to the contacting surfaces of the cable and the socket. These surfaces should be a abraded with steel wool, also coated with compound. (This method removes the oxide film, while the vaseline inhibits the formation of a fresh oxide film.). The crimping or the bolting of the socket should be done immediately thereafter. The compound will get squeezed out from the true contact points, and any excess of compound will prevent oxidation.
- Threaded fasteners should be re-tightened periodically, the periodicity being determined by the measurement of temperature on a few pilot sockets.

Since so much depends on the quality of the crimping tools and the sockets used on the shop floor, special steps have to be taken on the shop floor.

- Every time the crimping tool and dies are checked and/or tested, a metal tag indicating the next date of checking should be attached to the tool.
- The staff should be instructed not to use the crimping tools and dies after the expiry of the date marked on the attached metal tags. Such tools should be returned to the tool room.
- Samples of correctly crimped sockets should be exhibited on suitable boards in the section of the shop floor where they are fitted. The staff should be advised to report any visible differences that may arise in the materials supplied to them.
- Small, hand operated crimping tools are usually fitted with ratchet locks—as seen in Fig. 6.4—which prevent partial crimping. The correct operation of these locks should be verified periodically by the supervisors.

Where large numbers of crimped joints are made every day (such as in production units) using small hand operated tools, the workers may

Failures of Plug/Socket Connectors

of the need for flexible, detachable connections. Similar couplers are used between coaches of Electric Multiple Units. Flexible couplers of smaller sizes and very low voltages are used for the interconnection of computer units. The three pin plugs and sockets used with domestic appliances are amongst the simplest of these connectors.

All these different types of connectors of widely varying current and voltage ratings have one common feature: each plug or pin is firmly gripped by a socket and the gripping force is developed by tiny springs which deflect when the plug is inserted into the socket. The initial provision and subsequent maintenance of this force is the key to the safe and reliable operation of these couplers. As it is not practicable to measure the gripping force directly, it is measured indirectly by measuring the frictional force encountered in pulling the plug out of the socket.

When it is necessary, on account of operational or maintenance needs, to disconnect and reconnect one or more wires between equipment from time to time, plugs and sockets are used instead of bolted connections. When a number of connections are to be made simultaneously, multicore cables and multicore plugs/sockets are used.

Figure 5.3 shows a cross section through a single plug and socket assembly. Many such assemblies are fitted side by side in insulated blocks and casings to form a multicore coupler as shown in Fig. 7.1.

A common type of multicore coupler used in electric locomotives is the 19 core coupler. It is used in a number of locations including:
- Couplers between locomotives
- Master controller

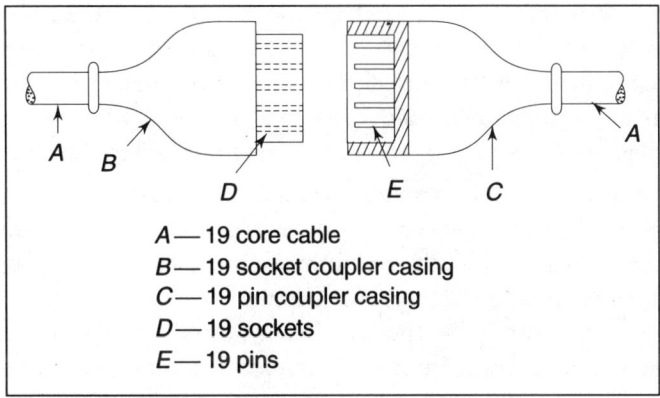

A — 19 core cable
B — 19 socket coupler casing
C — 19 pin coupler casing
D — 19 sockets
E — 19 pins

Fig. 7.1 Multicore coupler for 19 core cable

- Drivers desk
- Control switch and fuse panel

Multicore couplers and multicore control cables are generally used for one or more of the following reasons:

- When a large number of control cables have to be laid over relatively long distances, cable laying becomes easier, faster and neater when multicore cables are used.
- Trouble-shooting becomes easier and faster when multicore couplers are used. It is easy to isolate equipment during the process of narrowing down the faulty zone.
- Maintainability is improved because complete subassemblies can be disconnected and taken out of the installation for overhaul or repairs. Unit exchange systems of maintenance can be adopted.
- The presence of the protective sheath on multicore cables minimises the chances of damage to and failures of control cables.

Another application where multicore cables and couplers are used is the interconnection between coaches of trains. In long-distance, main line trains, the interconnectors are for transmission of power for train lighting and air-conditioning. In Electric Multiple Unit stock in use on the suburban sections of metropolises like Bombay, Calcutta etc., the electrical couplers are used for the transmission of control signals between the driver and the driving motor-coaches. Control wires for the electro-pneumatic brakes are also amongst the wires in these multicore cables (see Fig. 7.2).

When a coach has to be removed from or added to a train formation for repairs or maintenance, these multicore couplers are very convenient. About 50 to 150 electrical connections can be disconnected and, more importantly, reconnected accurately within minutes. If even two wires were to get interchanged, as might happen if they were all connected individually by hand, the train could become inoperative. Some electrical equipment could also get damaged. Use of multicore couplers avoids all such problems apart from saving time.

Users of personal computers would be familiar with the multicore couplers between the central processing unit (CPU) and the video monitor, the keyboard, the printer, the mouse etc. Similar couplers are used inside the CPU for interconnection between cards, drives and power supply modules. The couplers as also the cables are smaller in size and the shapes are also different; but the basic principles of design, maintenance and testing are exactly the same. Defects in these couplers are unlikely to lead to fires because the currents and voltages

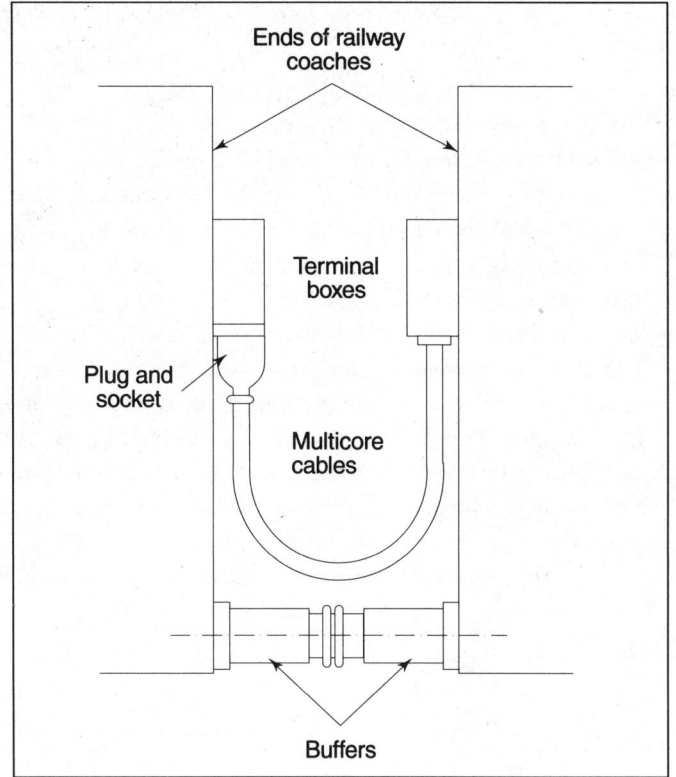

Fig. 7.2 Multicore control cables between railway coaches

are very small, but malfunctioning of computers and electronic devices is very often due to defects in multicore couplers.

A very common type of multicore coupler in everyday use in every home or office is 125-V and 250-V plugs and receptacles used for connecting electrical lamps and appliances. Mating plugs and receptacles are manufactured in many different voltage and current ratings for different applications. The National Electrical Manufacturers Association (NEMA) publishes standards and configuration charts for these wiring devices.

7.2 SPECIFICATION FOR PLUGS/SOCKETS

There are hundreds of types of plugs and sockets which vary with regard to the number of pins, current ratings, voltage ratings, shape and size. These are covered by various standards, dealing with similar aspects. We shall discuss one type here viz., IS 1293 for 6 A/16 A, 250-V plug/socket.

This standard gives the following important details:
- Pin diameters with their tolerances
- Pin and socket spacing with their tolerances
- Temperature rise at rated current (not to exceed 20°C)
- Current breaking capacity (125 percent rated current at 110 percent rated voltage. Test 10 times at 30 second intervals. Millivolt drop not to exceed 50 after test.).
- *Pull-out force test:* For this test, the socket is held upside down, a test pin of specified dimensions and weight is inserted in the socket and released. The test pin shall not slide out under its own weight. This test requirement confirms that adequate contact force is provided between the socket and the pin. As stressed repeatedly before, this contact force is the most important parameter for ensuring reliability of the coupler. Figure 7.3 shows the test pin for the current carrying socket of the 16-A unit.

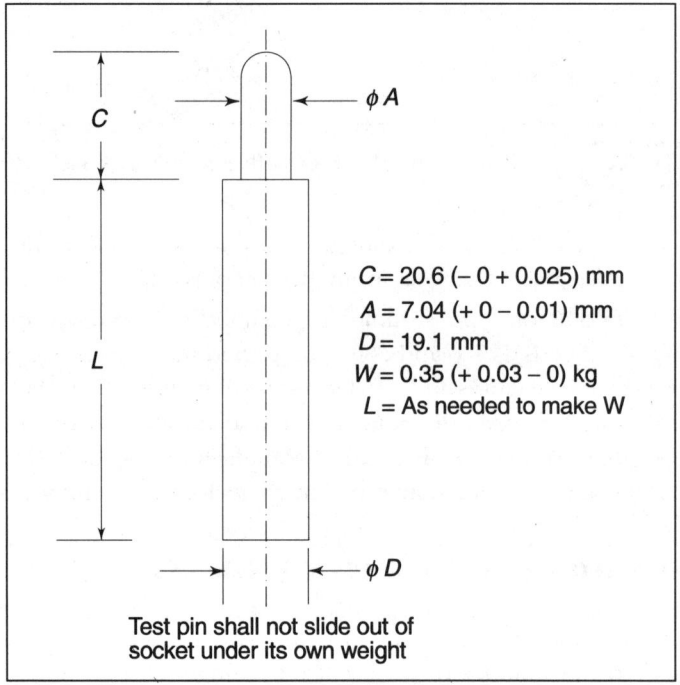

Fig. 7.3 Test pin for gravity pull-out test on socket

7.3 IMPORTANCE OF THE CONTACT FORCE

The force required for obtaining good electrical contact is provided by small springs inside the sockets. Initial provision and subsequent maintenance of adequate contact force between the plugs and the sockets is the key to the achievement of safe and reliable operation of these couplers. The greater the gripping force, the lower is the contact resistance. The coupler is so designed that the heat developed by the passage of current through its contact resistance does not increase the temperature of the assembly by more than about 25°C. Figure 7.4 shows two different spring arrangements usually provided for developing the required contact force.

Since the temperature rise limit of 25°C applies to all types of couplers, it is evident that couplers for carrying higher currents must have lower contact resistance and consequently, higher contact force.

If there are errors in the dimensions of the mating plugs and sockets, or if the springs are defective, the contact force between the plugs and sockets will be less than the required optimum level. Depending upon the extent of shortfall in the force, different failure modes manifest themselves.

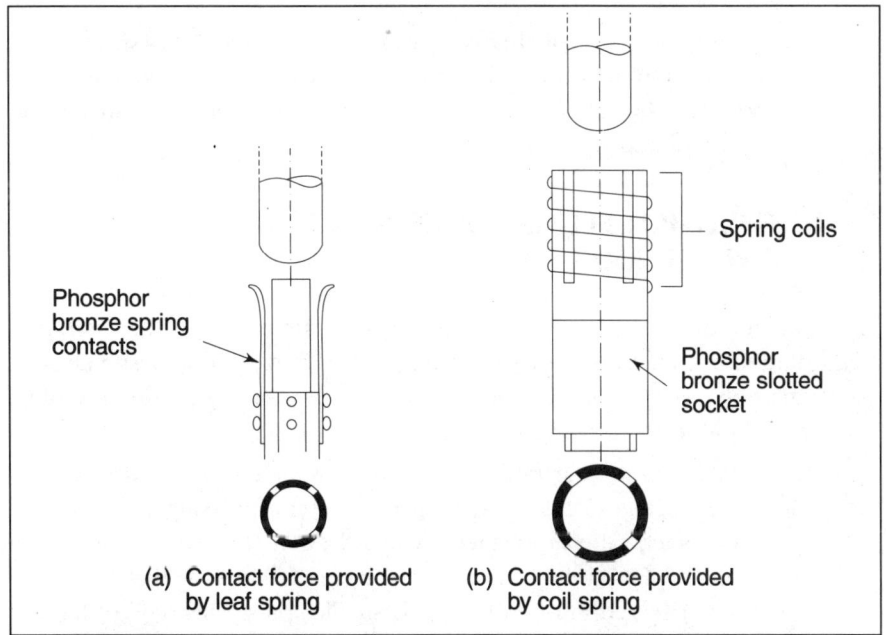

Fig. 7.4 Two different spring arrangements for multicore sockets

- If the contact force is very low or zero, the contact faces may get oxidised or coated with dust. In such cases, the contacts show an open circuit and this may cause some operational interruption if the contacts are in the control circuit. If the defective contacts are in the power circuit, an arc may develop at the bad contact and this could start a fire.
- If the contact force is lower than the specified minimum, but not zero, there is no open circuit; but the contact resistance would be high. The passage of current through this high contact resistance would generate excess heat at this point and the temperature of the connector would eventually go far beyond the normal temperature of about 70°C. The temperature may reach values at which the surrounding insulating material get carbonized and ignite.

What happens next after arcing or overheating in the connector, is largely a matter of chance. If a short circuit develops quickly, and some protective device operates, there would be simply a failure.

On the other hand, if the arc continues to flare and the fire spreads to the cables and structures, a major fire may be the result. This is the less likely result but the consequences may include considerable damage to equipment.

Failures and fires of this type can be prevented by taking a few simple precautions during the inspection at the stage of purchase of the couplers and also at the stage of installation and maintenance of these components.

7.4 THREE-PIN PLUGS/SOCKETS IN DOMESTIC AND GENERAL USE

A common example of a multicore plug/socket contact is the three-pin plug and socket used for supplying electric power to domestic electrical appliances. Figure 7.5 shows a typical domestic 15-A three-pin plug and wall-mounted socket.

Discolouration, charring, sparking etc. are often observed in three-pin plugs/sockets used for immersion heaters, pressing irons, room heaters or such other appliances which draw high currents. Since the larger earth pin rarely carries any current, failures of this type are generally observed in one of the two smaller pins. This type of failure is certain if a current much higher than the rating of the socket is drawn through it. However, similar failures are possible even when

Failures of Plug/Socket Connectors   129

the currents are well within the ratings of the sockets. This can happen due to either inadequate initial contact force between the sockets and the pins, or shrinkage during service of the plastic base of the three-pin plug.

The first of the two possible causes mentioned above, i.e., inadequate force between the pin and the socket, could be either due to errors in the dimensions of the pin or the socket, or due to a defective spring. If correctly designed, either the pin or the socket should be made of springy material and split longitudinally. The mating part is solid. The bore of the socket should be slightly smaller than the diameter of the pin so that, when pressed together, the springy component deflects slightly thereby providing the essential contact force.

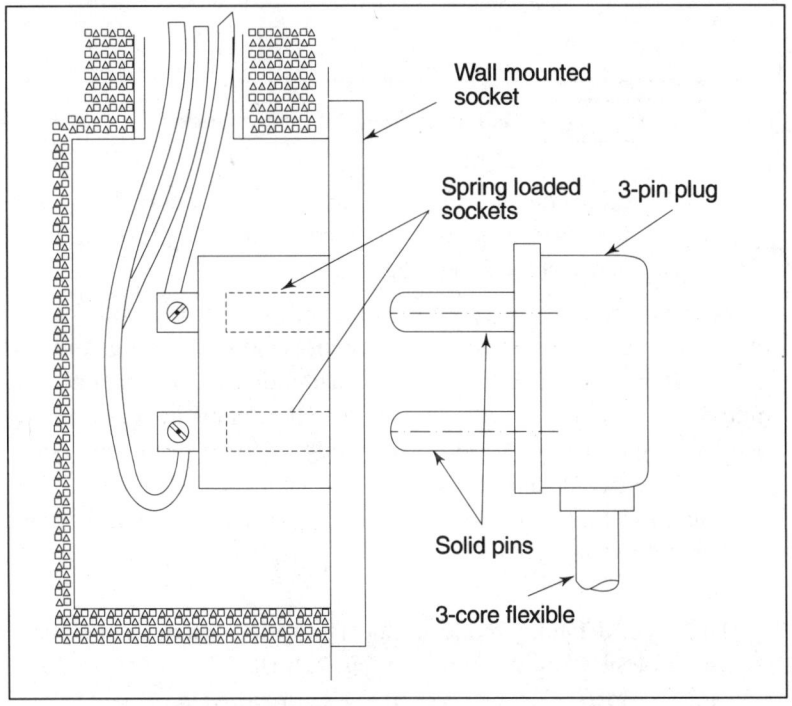

Fig. 7.5   **Domestic wall mounted socket/plug**

If the interference between the pin and the socket is not sufficient, the contact force between them will also be inadequate and this will lead to excessive contact resistance and overheating. Temperatures in excess of 300°C can be reached; the plastic material surrounding the pin or the socket will in such cases get charred, and it may even catch

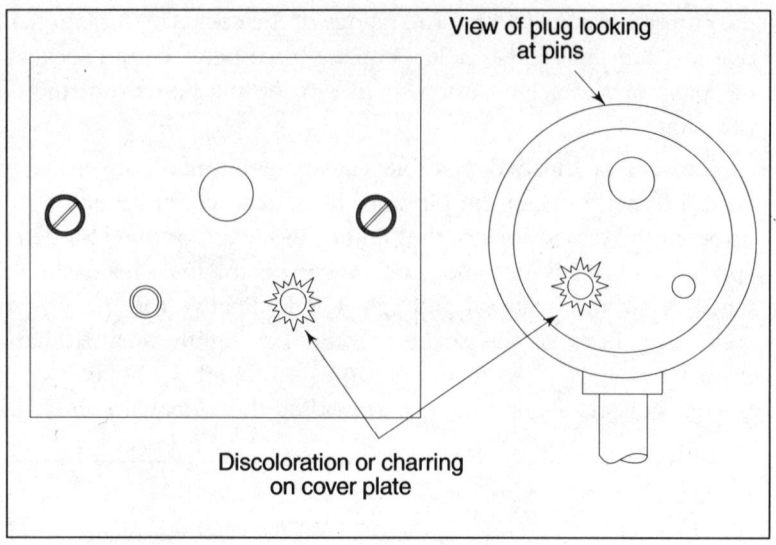

Fig. 7.6 Failure of domestic plug or socket

fire if allowed to remain in service. Discolouration and charring of the plastic base or cover around the defective pin and socket is the usual first sign of such defects (see Fig. 7.6).

Special care must be taken to use plugs/sockets of good design and manufacture, specially for applications where the plug and socket generally remains inaccessible and unattended as, for example, for an automatic geyser or air-conditioner which remains in continuous service. As a measure of caution, the pull-out force from the socket for each of the two smaller pins should be checked individually after dismantling the three-pin plug. Better still, a semi-permanent connection with a bolted terminal board may be used in place of a plug and socket.

The second failure mode is due to the loosening of the threaded terminal block from the pin [see Fig. 7.7(a)].

There is a vulnerable contact point inside the threads between these two components. If the pin is screwed on tight, there is no problem. If the threaded joint between the pin and the terminal becomes loose, the contact resistance rises and overheating is sure to follow. It is important that the pin should be fully tightened on to the terminal block initially during installation. This simple precaution is not taken by many electricians mainly because of the lack of awareness of a simple but necessary precaution.

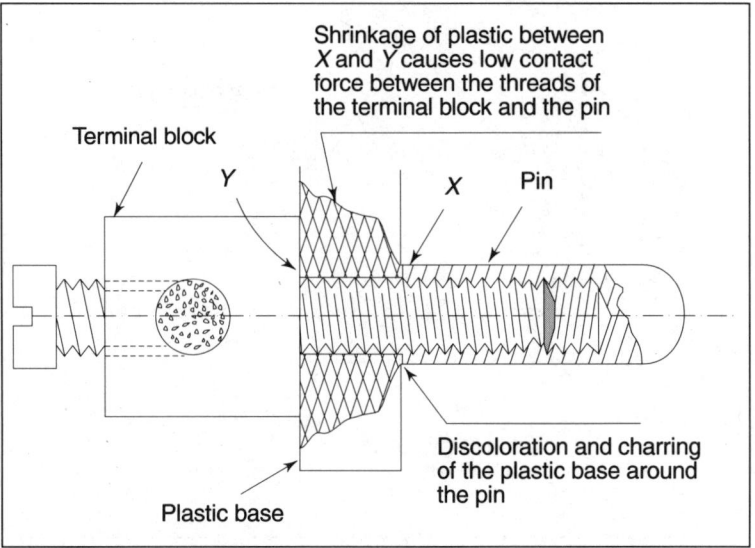

Fig. 7.7(a) Overheating of threaded joint between terminal and pin

Initial tightening of the pin onto the terminal block is not sufficient to ensure that there is no failure. The design of most plugs has a built-in defect. The thin bit of plastic [X-Y in Fig. 7.7(a)] between the pin and the terminal block will shrink in service and this will create a situation as if the pin was not tightened onto the terminal block fully. It is necessary therefore to tighten the pin periodically onto the terminal block. A maintenance-free design would provide a metal to metal tightening of the pin and the terminal block without any plastic material in between. This failure mode and the correct design for a terminal block is discussed in greater detail in Chapter 10.

A very reliable design of pin and socket shown in Fig. 7.7(b) rarely seen in India, avoids the type of problem described in the preceding paragraph. In this design, the threaded joint between the terminal block and the pin is eliminated and there is no interleaved plastic layer which could shrink and cause loss of contact force, overheating, charring and perhaps even fire.

7.5 INDUSTRIAL MULTICORE COUPLERS

Similar but larger fittings are used in heavy industry, railway locomotives and coaches, mainly for control cables and occasionally for power cables. One example of a multicore cable connector with plug/

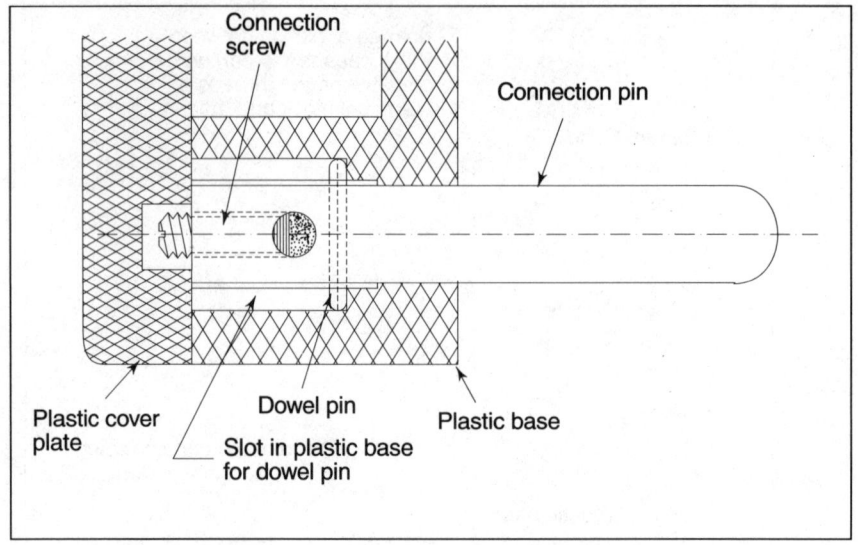

Fig. 7.7(b) Design detail of a reliable 3-pin plug

socket terminations is shown in Fig. 7.1. Details of the individual plug/sockets are shown in Fig. 5.3.

There are many different designs of multicore plugs and sockets, but they all have a few common features. The 'pins' and the 'sockets' are connected to insulated stranded wires through crimped joints. The sockets have spring-loaded contact segments. The free bore of the socket is smaller than the diameter of the plug pin by an amount which is carefully controlled. This is done by specifying certain tolerances on the dimensions of the mating components. When the plug and socket are pressed together, the springs around the contact segments have to bend or deflect, and this provides the force which is so essential for obtaining good electrical contact.

7.6 FAILURE MODES

Plug and socket connectors of the type described above are subject to failures of the following types:
- Overheating or burning of the contacts between the pins and sockets may be the initial failure mode. This may result in charring of the insulating material between the pins or sockets and even short circuits between adjacent pins/sockets.

Failures of Plug/Socket Connectors

- Short circuits between adjacent pins/sockets in the coupler could be the initial failure mode in some cases. The short circuit may either be due to tracking across the surface, or puncture through the insulating material between the pins or sockets. This failure mode is relatively rarer, as the voltages in couplers of this type are usually too low to cause such failures. However, failures of this type may occur due to very poor quality of the insulating material or entry of water into the space between the pins and sockets.
- Open circuits may occur due to excessive errors in the dimensions of the pins and the couplers or due to defective crimping of the pins and the sockets on the cables. The wires may get pulled out of the sockets due to inadequate crimping. Open circuits may also occur due to several other possible reasons, but such other defects are rare.

It is very important to distinguish clearly between the the first and the second failure modes referred to above. The failure mechanisms and the remedial measures to be taken in the two cases are quite different, although the appearance of the failed components may be very similar.

When a multicore coupler fails, the failed couplers should carefully be opened out, piece by piece, while observing and recording the observed features of the nature and location of damages. An effort should be made to determine the location of the first stage of the failure. The electrical position in the circuitry of the wires affected should also be noted.

7.7 OVERHEATING OF PIN/SOCKET

If, for any reason, the contact force exerted by the segments of the female contact on the pin of the male contact gets reduced, the contact resistance rises in inverse proportion. If the normal current through the contact is equal to its carrying capacity, the temperature may rise to excessive levels and this may, in turn, establish the vicious circle of increasing oxidation, increasing contact resistance, and increasing temperature. If such a circle is established, failure is only a matter of time. Sooner or later, the temperature will rise to such an extent, that it may cause charring and electrical short circuit through the insulating material, if not melting of the contacts.

Coupler failure by the mechanism described above, can only be prevented by ensuring that the initial contact force is sufficiently high and the initial contact resistance sufficiently low to limit the temperature of the contacting surfaces to less than 70°C. The actual value of the contact force to be adopted is a decision to be taken by the designer of the equipment and verified during type tests. The limits for contact force depend on the current to be carried by the contacts, the size of the couplers, and the vibration levels. These limits are usually given in specifications/drawings of the couplers.

It is not possible to measure the contact force directly. It is estimated indirectly by measuring the pull-out force and in fact, the limits of contact force are usually specified in terms of the pull-out force. The pull-out force is usually 25 to 30 percent of the contact force, the ratio being the coefficient of friction.

The pull-out force has to be verified separately for each socket using a test pin of diameter which is the lowest in the permissible range for the diameter of the pin. The weight of the pin should be equal to the minimum acceptable pull-out force. As in the case of the three-pin domestic plug, the coupler head may be held upside down and the test pin inserted in each socket in turn to confirm if the pin/socket are made of the required material and to the specified dimensions. No useful purpose is served by measuring the pull-out force for the complete multicore coupler. The diameters of the pins in a coupler should be measured with a micrometer and compared with the permissible range specified in the drawing for the coupler.

The measurement of pull-out forces with a test pin and the measurement of diameters of the coupler pins has to be done during acceptance testing of couplers. Similar measurements should be made when investigating coupler failures involving overheating of contacts and charring/burning of the insulation between the pins/sockets. Since it is often not possible to make these measurements on the failed contacts, they should be made on other contacts which have not failed in the same coupler or in other couplers of the same make and design. If during such measurements on other contacts, some contacts with pull-out forces which are significantly lower than the permissible lower limit are discovered, it is reasonable to conclude that the failure was due to overheating caused by inadequate contact force and excessive contact resistance. On the other hand, if the pull-out force measurements are all within the permissible range, the possibility of the coupler failure having been caused by insulation failure has to be considered.

Acceptance tests on couplers of this type should also include endurance tests, during which the couplers are repeatedly coupled and uncoupled until the number of such operations is at least double the number that may be expected to occur during the expected lifetime of the couplers. The pull-out forces, as also the millivolt drop across the contacts when carrying the rated current, are measured before and after the endurance test. The degradation in these parameters should not exceed the limits indicated in the specifications.

If the pull-out forces are less than the specified limits, there is obviously a defect in the design or manufacture of the socket. This has to be corrected by the manufacturer.

Overheating of either the pins or the sockets is also possible as a result of defective crimped joints between the pins or sockets and the stranded wires (see Chapter 6).

7.8 INSULATION FAILURES BETWEEN ADJACENT PINS/SOCKETS

Insulation failures have to be considered when the probability of overheating of the contacts due to bad contacts has been eliminated by measurements of pull-out forces and the millivolt drop tests at rated current. If there are any sockets which are unaffected by the failure of a few sockets the insulation resistance and breakdown voltage between adjacent pins or sockets should be measured. The possibility of water entering the coupler, in the case of outdoor couplers, may also be considered.

Acceptance tests should include type tests to measure the breakdown voltage between adjacent segments, and routine tests such as proof tests at voltages of the order of 300 V between adjacent segments for one minute.

7.9 OPEN CIRCUITS

Although failures of this type are relatively rarer, one possible cause is the presence of a clearance between the socket and the pin instead of the required interference. This condition is obviously due to very poor quality of manufacture. It can be detected very easily by the pull-out force measurements. The pull-out force would be almost zero. In such cases, overheating or burning does not occur, because the current carried through the pin/socket is zero.

Another cause of open circuits is poor crimping of either the socket or the pin onto the stranded wire in the multicore cable. This type of defect has been dealt with in detail in Chapter 4 which is fully applicable to the crimped joints between the stranded wires and the pins or sockets of multicore couplers.

7.10 MULTICORE CONNECTORS IN ELECTRONIC DEVICES

The currents and voltages used in computers, television sets and electronic equipment in general are much smaller than those used in industrial power equipment. However, the basic principle—that a certain minimum force between the plug and the socket must be provided to ensure good electrical contact—applies to all plugs and sockets, including those in electronic devices.

Thin oxide films on copper conductors are unavoidable. These thin films do not create any problems at voltages of the order of 240 or 480 in industrial equipment, because the films get punctured at the operating voltages. In electronic circuits, the operating voltages are of the order of 5 V or even less. At these low voltages, the copper oxide films remain intact and result in high contact resistance. Hence, in electronic devices, the connector contact surfaces are usually given a gold plating of thickness of approximately two microns. Gold does not oxidize and is also not affected (unlike silver) by sulphur compounds in the air. It maintains good electrical contact even when the voltages are low. Here too, the contact force has to be controlled within certain limits.

7.11 OPEN CIRCUITS DUE TO EXTERNAL FORCES

The wires may get pulled out of the crimped joints with the pins or sockets due to external forces on the multicore cables. To prevent this from happening, it is necessary to clamp the cable to the casing as shown in Fig. 7.8.

It is necessary also to leave a little slack in the individual cores inside the coupler heads. Moreover, the clamping should be done after the cable is bent to its final position in order to prevent tension in the outer cores around the bends. If this is not done, the copper strands in the outer cores are liable to break due to fatigue.

Failures of Plug/Socket Connectors

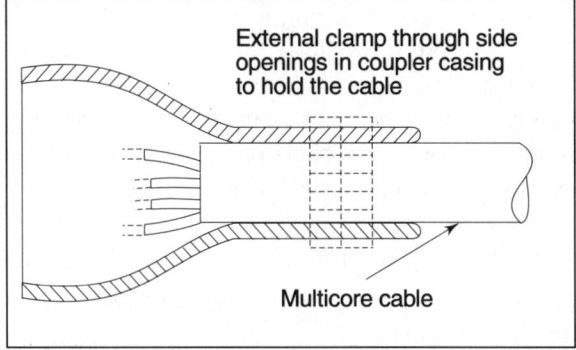

Fig. 7.8 Clamping of multicore cable to prevent failure

7.12 PRECAUTIONS DURING INSTALLATION OF MULTICORE CABLES

Crimping the sockets or pins onto the cores is very important (see Chapter 6 on crimped joint failures). The pins should be inspected to ensure that the outer diameters of the pins are within the specified limits. Go and No-go gauges may be used for this purpose.

The sockets should be inspected by carrying out the pull-out test on a specially designed test pin. This test pin should be of a diameter equal to the smallest permissible diameter of the pin, and should also have a weight as specified. The coupler is held upside down and the test pin is inserted into the socket under test and released. The pin should not slide out under its own weight. It is not necessary to test every socket on every multicore socket. Acceptance sampling methods may be used as specified.

7.13 SUMMARY

Multicore coupler socket failures can occur in different failure modes due to a variety of seed-defects. As the final appearance of the failed components is identical for different failure mechanisms, it is necessary to examine not only the failed components, but also the components which have not failed in the same coupler or in other couplers of the same make, type and age.

The common defects and seed-defects in multicore couplers are summarised below:
- Low pull-out force due to:

- weak or broken spring in socket
- socket bore too high
- pin diameter too low
- Tinning or silver-plating of pins or sockets worn
- Crimping of pins or sockets on wires poor
- Defective insulation between pins/sockets due to
  - entry of water
  - poor quality of material

The following items must be checked at the stage of acceptance of multicore plugs/sockets;
- Pin diameters
- Pull-out forces of individual core sockets
- High voltage proof tests between pins and between sockets.

7.14 DO'S AND DON'TS FOR PREVENTING FAILURES OF MULTICORE COUPLERS

- Check the pull-out force (i.e. the drop-out weight with the test pin) individually on all sockets, using the appropriate test pin when:
  - accepting new sockets/plugs
  - in the event of any failures
- Check the pin diameters with mircrometers or Go/No-go gauges when accepting new plugs.
- Carry out hipot proof tests between adjacent sockets and pins when accepting new couplers and when investigating failures of plugs/sockets.
- Clamp the multicore cable in the coupler head after the cable is bent in its final position.

Chapter 8

Fractures of Crimped Sockets

In this chapter, we will discuss:

- Failures of electrical connectors due to fracture of crimped sockets.

- Special conditions under which crimped sockets are prone to fracture.

- Failure modes and failure mechanisms involved in crimped socket fractures.

- Metal fatigue, the process which leads to cracks and eventually to fractures.

- Factors which initiate and accelerate metal fatigue: vibration, thermal cycling and stress raisers.

- Measures and precautions to be taken in design, manufacture and installation to prevent fractures of crimped sockets.

8.1 INTRODUCTION

There are five major causes of this mode of failure. These are:
- Defective material.
- Sharp bend or stress raiser introduced during manufacture of the socket.
- Stress raiser introduced as a result of defective crimp.
- Excessive vibratory flexing of the wire.
- Thermal stresses due to alternate heating and cooling of conductor under constraint.

The mechanism of failure in all cases of this mode of failure is usually the same, viz. metal fatigue, crack formation, crack propagation and finally, fracture. If the current and voltage are low, there may be a clean break and the small arc may extinguish itself. There would be some kind of operational interruption, but there would be no fire.

On the other hand, if the current and voltage are high enough, the arc which is formed when the fracture occurs would continue to flare. If the arc gets transferred to the return wire or to earthed structures, there would be a sudden surge in current, the protective system would operate and the arc would be extinguished. Again, there would be a failure but no fire.

If the arc does not get transferred to the return wire or to an earthed structure or if the protective system is defective, the arc would continue to flare for a long time. Arc temperatures are generally of the order of several thousand degrees Celsius. At such temperatures, combustible material in the vicinity would catch fire due to the heat radiated by the arc. In such a case, a major fire is likely to start within a few minutes.

In this chapter, the factors which accelerate the fatigue process which ends in the fracture of the socket, as well as the measures that should be taken to prevent fractures of crimped sockets are described.

Fractures of crimped sockets are rare in stationary installations. They occur occasionally in electric locomotives and electric multiple unit stock. They may also occur in machines or equipment in stationary installations if the machine itself is subject to severe vibrations or if the cable is connected to a moving or portable component.

As in other failure modes, the true cause of failure of crimped sockets, which could be fatigue fractures of the type described in this chapter, may not be identified as such, since arcing and fire may burn off the crucial evidence.

8.2 FRACTURES OF CRIMPED SOCKETS DUE TO MATERIAL DEFECTS

Crimped sockets are generally manufactured out of electrolytic or high conductivity copper tubes. The sockets are fully annealed so as to ensure that the severe deformation and compression which has necessarily to be effected during manufacture and during crimping, does not produce any cracks in the socket. Such cracks act as stress raisers.

Fractures of Crimped Sockets

If the cracks are in a zone where there are high alternating or fluctuating stresses during service, the cracks will grow and ultimately cause fracture. Therefore, the ultimate tensile strength, the yield point and the elongation must all be within the limits specified for annealed electrolytic copper.

Another effect of defective material could be in the form of reduced conductivity of the copper (see Chapter 6).

The quality of the material of the socket can be checked easily with the help of a conductivity meter. This meter consists of an electronic device with a probe which is to be placed in contact with the material to be tested. The meter works on the principle of eddy currents induced by a coil in the probe. The meter gives a direct reading of the conductivity percentage of the material. If the conductivity is above 99.9 percent, it is safe to assume that the material is acceptable.

The extent of annealing can be determined by measuring the Brinell hardness of the barrel of the socket. This test is also very simple and practical even on the shop floor.

Another simple but critical test which can be made on the shop floor is the *bend test*. In this test, the barrel of the socket is fully flattened in a press. The test requirement is that there should be no cracks on the edges of the flattened piece. A magnifying glass should be used to check for cracks (see Fig. 8.1). As this is a destructive test, it can be carried out only on random samples drawn from the lot.

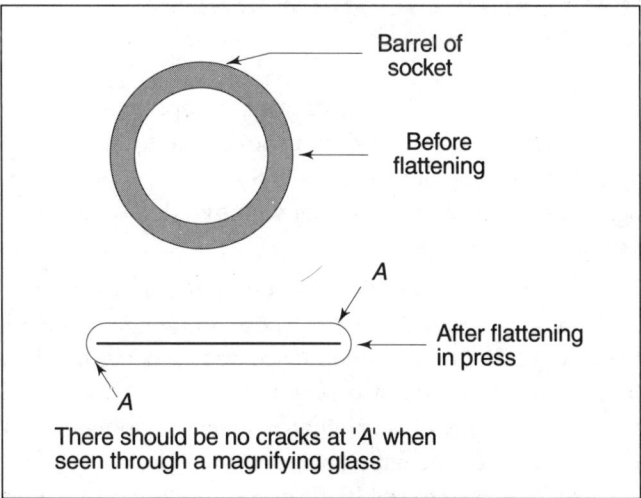

Fig. 8.1 Flattening test on socket barrels to check copper quality

It is normally not necessary to check the tensile strength and yield point if the conductivity, Brinell hardness and bend test results are satisfactory.

Material defects will also be exposed by careful examination of the edges of the palm of a socket where it has been flattened. However, the absence of cracks is not conclusive, because hairline cracks can get filled up by the tinning. Hence it is desirable to do the destructive test.

Even though the copper originally used for making the sockets is of the required quality, it is possible that it might get damaged during manufacture. Usually, copper has some oxygen dissolved in it. When such oxygen-bearing copper is heated (for annealing, for instance) in an atmosphere containing even small quantities of hydrogen, it gets embrittled, and cracks which are initially invisible can develop. Therefore, care has to be taken during manufacture to ensure that oxygen-bearing copper is heated in an atmosphere which is free from hydrogen.

Material defects are rare. In fact, defects of any kind should normally be expected to be rare; but this is no reason for waiving tests at the appropriate stage. It is often the rare and unexpected appearance of defects which is responsible for all failures in service. Zero failure performance can be achieved, but only by exercising constant vigilance at every stage.

8.3 FRACTURE DUE TO SHARP BENDS AND STRESS RAISERS

Even when the material is satisfactory, it is possible for cracks to start from stress raisers introduced in a socket due to either a design defect or a manufacturing defect. Figure 8.2 shows a socket with such a defect and Fig. 8.3 shows a socket with two defects: sharp corner at zz and excessive crimp at yy.

Fractures of crimped sockets generally occur along lines xx, yy and zz as shown in Figs. 8.2 and 8.3. These are generally fatigue fractures. The subject of fatigue has been discussed in detail in Chapter 13, but the important points are also stated here.
- Fatigue fractures occur when a metal component is subjected to alternating or fluctuating stresses if the peak tensile stress in each cycle of stress exceeds a certain limiting value known as the

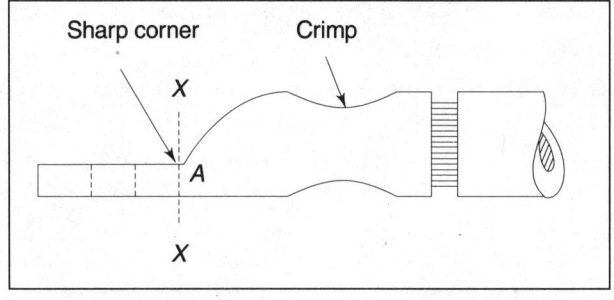

Fig. 8.2 Stress raiser (sharp corner) in socket

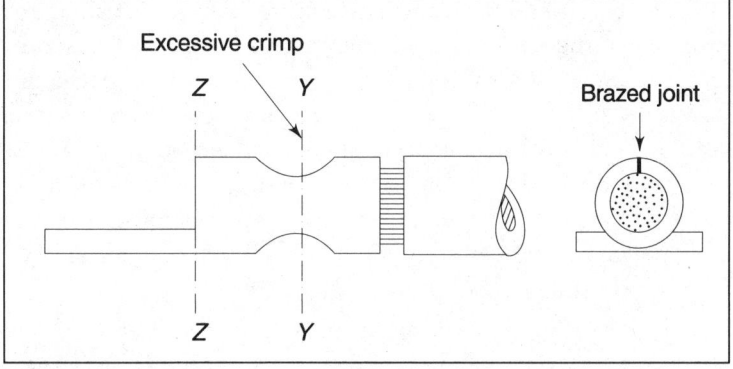

Fig. 8.3 Sharp corner and excessive crimp in crimped socket

endurance limit of the metal. For copper, this endurance limit is rather low, i.e., about 69 MPa or about 7 kg/mm^2 at the conventional number of 500 million cycles.
- Wherever there is a sharp change in section of a component, the stress on the surface is increased by a multiplying factor known as the *stress concentration factor* applied to the average stress. This factor depends on many dimensional parameters, the most important of which is the radius of curvature at the change of section. For small radii, this stress concentration factor could be extremely high.
- The radius of curvature at point A (Fig. 8.2) of the crimped socket should not be less than twice the tube wall thickness. A sharper bend is likely to produce micro-cracks on the outer surface of the bend. Cracks produced during manufacture have a very high stress concentration factor. Often, these cracks are hairline cracks which are not visible to the naked eye.

- There is acute stress concentration at the point A in Fig. 8.2 due to either bending or cracks.

It would be clear from the above that fatigue cracks are very likely to start and grow rapidly from defects such as those shown in Fig. 8.2. The remedy is obvious. Crimped sockets must not have any of the defects shown in Fig. 8.2. The sockets should be of the shape shown in Fig. 6.1.

8.4 FRACTURES DUE TO CRACKS FORMED BY DEFECTIVE CRIMP

The process of crimping involves considerable deformation and compression of the socket. Cracks may be produced in the socket if the crimp is excessive or if the crimping dies have sharp edges. It should be ensured that the shape of the dies provides a smooth change of section, so that the maximum crimp is obtained in the centre and there is a graduated reduction in section along the axis. There must not be any sharp edges or corners on the dies. All these precautions are generally taken by the manufacturers of the crimping tools and dies; but it is wise not to take this for granted.

Use of a crimping tool meant for a smaller size of socket can also result in the development of hairline cracks at the crimp location, and these may develop into fractures. Therefore, care should be taken to use the tool appropriate to the cable size.

The crimping die design and process should involve compression towards the axis of the barrel, and there should be no shearing or pinching of the barrel.

The mechanism of failure is the same as in the case of cracks or sharp bends, and all other precautions mentioned there are applicable in this case too.

8.5 FRACTURES DUE TO EXCESSIVE VIBRATION

As stated earlier, fatigue fractures occur when there are alternating or fluctuating stresses. Vibration is the most common cause of such stresses. The best method would obviously be to eliminate vibration but this is not always practicable. For instance, in electric rolling stock or locomotives, there is a limit below which it is not possible to reduce the vibration, although all efforts are made in that direction. Similarly, in certain industrial machines vibration or even some reciprocating

Fractures of Crimped Sockets

movement is unavoidable. Figure 8.4(a) shows how a crimped socket fixed to a vibrating base with a fairly long unsupported cable could be subject to alternating stresses due to the vibratory flexing of the cable. Figure 8.4(b) shows what happens when there is relative movement between the connected components and not merely vibration of the cable.

If, in the arrangement shown in Figs. 8.4(a) and 8.4(b), the cable vibrates as shown, an alternating tensile and compressive stress is produced at A. The palm of the socket is rigidly held by the screw and the barrel of the socket is very rigid compared to the cross section of the palm at A. The flexing of the socket takes place mostly at A and the stresses are high. Failure by fatigue, sooner or later, is the inevitable result.

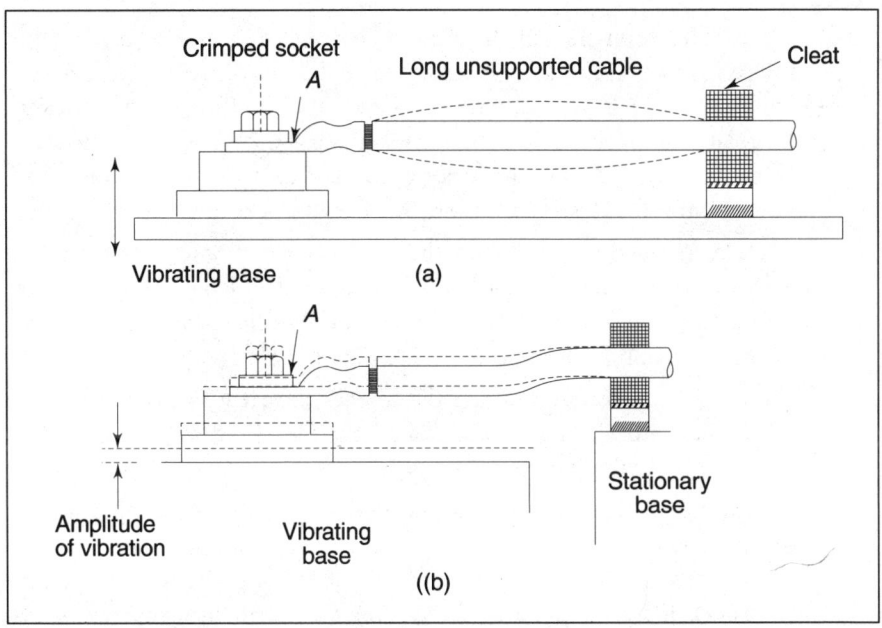

Fig. 8.4 (a) Failure of crimped socket due to cable vibration (b) Failure of crimped socket due to vibration of terminal board

It is therefore necessary to minimise the effects of vibration and movements by taking suitable measures. The basic principle of the measures to be taken is to eliminate the vibration at the socket and to transfer it to the stranded cable. This is done by clamping the cable on the vibrating component as shown in Fig. 8.5.

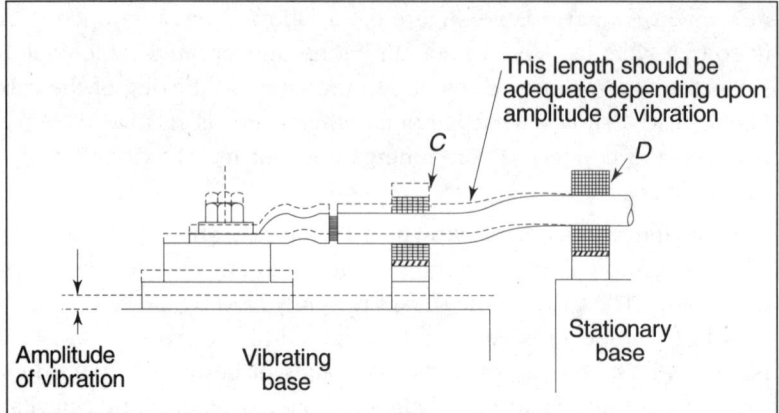

| Fig. 8.5 | Prevention of crimped socket failure despite terminal board vibration |

It will be seen that in the arrangement shown in Fig. 8.4(b), the vibration affects the socket, whereas in the improved arrangement shown in Fig. 8.5, the vibration has no flexing effect on the socket. The relative movement or vibration is absorbed by the cable between the cleats C and D. Adequate length of cable must be provided between C and D, depending upon the amplitude of the vibration.

While fixing the cable(s) in the cleats, care should, however, be taken to ensure that the cables are held firmly in the cleats but not pressed too hard, and that the edges of the cleats are rounded to prevent damaging the insulation (see Chapter 10).

It should be ensured that the distance of the cleat from the stripped end of the wire is long enough to prevent tracking failures over the surface of the cable (see Chapter 18).

It must also be noted that a stranded wire can absorb vibration at right angles to its length, but not along it. Therefore, it is often necessary to introduce a loop in the cable between the cleats as shown in Fig. 8.6(a). If the amplitude of the vibration is small, it may be sufficient to provide only a half-loop or even a quarter-loop as shown in Figs. 8.6(b) and 8.6(c).

8.6 FRACTURES OF SOCKETS DUE TO THERMAL STRESSES

Electrical cables get heated up due to the passage of current. The heat developed in, and consequently the temperature rise of the cable, is proportional to the square of the current.

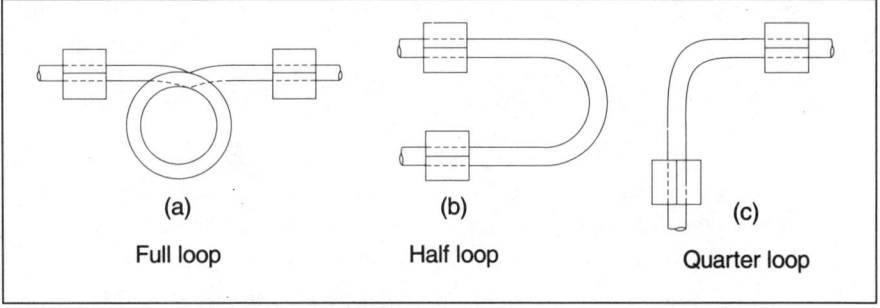

Fig. 8.6 (a) Loop in cable to absorb vibration along cable direction (b) Half-loop in cable (c) Quarter loop in cable

As the current in the cable is often subject to wide fluctuations depending upon the load, the temperature of the cable rises and falls. This causes the copper to expand and contract alternately. If the copper were prevented from doing so by mechanical constraint, thermal stresses are produced in the copper. The magnitude of the stress is higher than the endurance limit of copper for a temperature fluctuation of only 40°C. Use of stranded cables prevents the development of thermal stresses, as the strands deflect radially by the very small amounts necessary to prevent development of thermal stresses.

If solid conductors such as busbars are used, this phenomenon can create problems. Therefore, if there are cases of fractures of solid conductors, this possibility should be looked into. Arrangements such as flexible expansion joints may have to be provided to allow expansion and contraction to take place without creating stresses on the conductors or terminations.

8.7 INVESTIGATION OF FRACTURES OF CRIMPED SOCKETS

It may be possible to detect a crack in the incipient stages before failure takes place in service if similar sockets are examined carefully. If a crimped socket failure is suspected where the visible evidence is only molten and congealed metal, the only possible way is to check other sockets in service for not only cracks but also for the design deficiencies described in this chapter and also in the chapters relating to the other failure modes.

Watching the sockets and cables while the machine is operating under various operating conditions may reveal vibrations and suggest possible measures such as clamping of cables in cleats to prevent the stresses on the sockets.

If a fracture of a socket is detected, as sometimes does happen in low voltage, low current circuits due to the self quenching of the arc, further steps should be taken to check the material of the socket, its shape and dimensions, and possible vibrations, thermal stresses etc.

If the magnitude of the peak stresses exceeds the endurance limit of the material, fatigue cracks will develop, and eventually, the socket will fracture. The peak stress depends on the following factors, which should be investigated:

- The amplitude of the alternating flexure of the socket. This in turn depends on the amplitude and frequency of vibration of the machine and the stiffness of the socket/cable system.
- The magnitude of the stress concentration factor at the point of maximum stress on the socket. This depends on the radius of curvature at the bends.
- The dimensions of the socket/cable system.

There may be no need to look into these aspects in an existing installation if there is no case of fracture of a crimped socket; but since very little time and effort is required to check some of the factors, it is desirable to do so. If and when such a failure occurs, or if there is a fire in which some crimped sockets have melted, then an examination of all the possible causes, mentioned in this and the following chapters, is essential.

If there is an electrical fire in an installation where crimped sockets are used, and if there are signs of molten cable terminations, it is desirable to check the cable terminations very carefully, not only in the affected installation but, also in similar installations, elsewhere in the same organization.

8.8 DO'S AND DON'TS FOR PREVENTING FRACTURES OF SOCKETS

Crimped sockets are intrinsically very reliable and can be expected to last without cracking or fracturing for the entire lifetime of the electrical equipment, i.e., for more than 40 years. However, such a performance can be obtained only if the following precautions are taken during the design, manufacture and installation of the sockets.

- Ensure that the socket is made of annealed electrolytic copper. Verify the electrical conductivity, the Brinell hardness and the bend test.
- Do not heat either the raw material or the socket in a reducing flame or reducing atmosphere, at any stage during manufacture, specially if the copper has some dissolved oxygen. Such heating causes hydrogen embrittlement.
- Ensure that the dimensions of the socket, in particular the thickness of the barrel, complies with the drawing.
- Ensure that the design of the socket and the design of the crimping tool are proven in endurance tests on finished joints.
- Ensure that the crimping dies do not have any sharp changes in section or sharp corners or edges.
- Ensure that no vibratory stresses are produced on the sockets, by clamping the cables in cleats in such a way that there is no relative movement between the cleat and the terminals where the sockets are bolted on.

8.9 CONCLUSION

If the precautions listed above are taken, there will be no possibility of fractures of crimped sockets. On the other hand, if one or more of these defects mentioned above are present, fracture of a socket is likely to occur sooner or later, depending upon the operating conditions. Every socket is not likely to fail, but a few are sure to do so.

Whether a particular socket with one or more of the seed-defects mentioned above will ever fail through fracture, depends upon a number of factors such as the magnitude of the vibration, presence or otherwise of cracks at the point of bending, the nature and magnitude of the impurities in the copper, etc. It is therefore possible that some sockets with the above mentioned defects may continue to remain in service without failure for years, but they will always remain vulnerable to sudden failure without warning. The only possible way by which we can be assured of reliable service throughout the life-time of the equipment is to ensure that the seed-defects mentioned above are not present.

Chapter 9

Failures of Wire Strands

In this chapter, we will discuss:

- Failures of electrical connectors due to fracture of wire strands.

- Similarities and differences between fractures of wire strands and crimped sockets.

- Special conditions under which wire strands are prone to fracture.

- Failure modes and failure mechanisms involved in wire strand fractures.

- Metal fatigue, the process which leads to fractures of wire strands.

- Factors which initiate and accelerate metal fatigue: vibration, thermal cycling and stress raisers.

- Measures and precautions to be taken in design, manufacture and installation to prevent fractures of wire strands.

9.1 INTRODUCTION

Failure of electrical connectors due to wire strand fracture is not a very common type of failure. Nevertheless, it is an important failure mode because failure can lead to fire and extensive damage.

Wire strand fractures generally do not occur on stationary installations, but they sometimes do occur on the electrical wiring of locomotives, coaches, electric multiple units and machines subject to heavy vibration. However, these failures occur only when certain seed-defects are present in the installation.

Failures of Wire Strands

The mechanisms of failure, the factors which initiate and sustain these mechanisms and the measures to be taken to prevent wire strand fractures are similar to those which occur on crimped sockets as discussed in detail in Chapter 8. These are, therefore, merely enumerated in this chapter.

Parting of overhead transmission and distribution lines are generally due to damage caused by electrical arcing resulting from bad contacts in joints and short circuits, but occasionally, the failure mechanisms described in this chapter become relevant.

There are a few causes and remedies which are applicable specifically to wire strand fractures. These are as follows and are discussed in detail in Sections 9.3, 9.4 and 9.5.

- Wire strand fractures due to vibration can be minimised by the use of heat shrinkable sleeves.
- Sharp edges on the inside of the socket barrel are sometimes responsible for wire strand fractures.
- Fatigue fractures of wire strands inside their insulating sleeves are due to alternating flexing of the multicore cables.

The period that would elapse between installation and failure depends on a number of factors such as the amplitude and frequency of the vibration, quality of the crimp and cable socket, the distance between the socket and the cleat, and the direction of the vibration. Some sockets with such seed-defects may fail within a few months, but some may go on for years without failure. Some may not fail at all. On the other hand, if we ensure that such seed-defects are not present in our installations, we can be sure that there would never be any failures of wire strands.

9.2 CAUSES COMMON TO SOCKET FRACTURES AND WIRE STRAND FRACTURES

The causes common to wire strand fractures and crimped socket fractures are:
- defective material
- excessive vibratory flexing of the wire

Although the above two factors are common, there are also a few differences. Material defects are likely to be very rare in the case of wire strand fractures because the process of wire drawing would eliminate many types of defective materials. The only possibility is of any damage caused in the final annealing process. However, this remote possibility may have to be considered if no other causes are discovered during investigation of wire strand fractures.

The other common cause, i.e., vibration, should be considered first when investigating wire strand fractures. The remedies to be applied are exactly the same as those recommended for preventing crimped socket fractures. These are:

- Provision of cleats to eliminate flexing at the socket and to transfer the flexing to a portion of the cable between two cleats.
- If the fracture location is at the crimp section, the possibility of excessive crimp or sharp edges on the crimping dies should be looked into. The remedy, obviously, is to modify the crimping dies.

The arrangement proposed in the first point above is the same as that suggested in Fig. 8.5 relating to fractures of crimped sockets due to similar effects of vibration. It is important to note that vibration of the complete assembly of the socket, the cable and the cable cleat does not lead to fracture of either the socket or the wire strands. It is the relative vibratory motion and the resultant flexing of the socket or the wire which is harmful.

By fixing the cable cleat on the vibrating machine, flexing of the wire is not eliminated. It is transferred to another part of the cable where it is unlikely to cause as much stress in the wire strands as when flexing takes place close to the socket. It is necessary, however, to ensure that there is adequate distance between the two cleats C and D in Fig. 8.5 where the relative motion is allowed to take place. In case there is regular movement and not merely vibration, it may become necessary to provide a loop in the cable between the two cleats.

9.3 HEAT SHRINKABLE SLEEVES

In the case of insulated wires of small cross-section (2 to 10 mm^2), it is often not practicable to clamp the wires on the vibrating machine. In such cases, there is another practical way to reduce the stress on the wire strands at the point where they emerge from the sockets. by providing a heat shrinkable sleeve which grips both the socket and the wire insulation (Fig. 9.1).

The sleeve provides electrical insulation over the barrel of the socket, but more importantly, it eliminates the mechanical weakness

Failures of Wire Strands

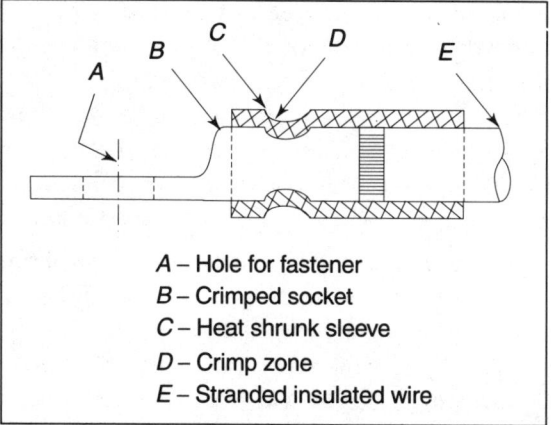

Fig. 9.1 Heat shrinkable sleeve to prevent strand fractures

which is introduced by the absence of insulation over the small portion of the bare conductor between the end of the socket and the end of the cut insulation of the cable. This prevents stress concentration in the exposed part of the insulated cable and consequent fracture of the strands due to metal fatigue (see Fig. 9.2).

This method is particularly suitable for control cables. They are small in size, but are as vulnerable to failure as any other cable and their failure can bring very large machines to a halt. It is not practicable to provide clamps on thousands of individual cables of small size, but it is very convenient to use heat shrinkable sleeves over them to improve their mechanical and electrical strength.

In the absence of such a sleeve, the flexing of the wire gets concentrated in the relatively more flexible bare copper zone between

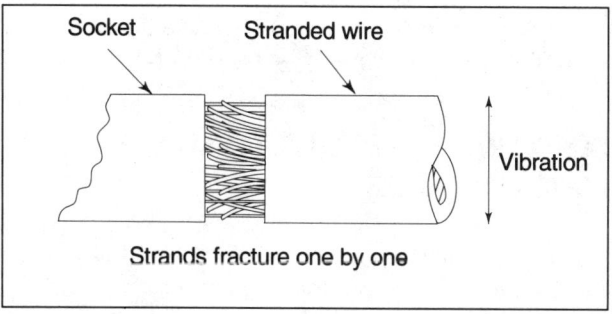

Fig. 9.2 Strand fractures between socket and cable insulation

the end of the socket and the end of the insulation as compared to the insulated wire. Provision of the sleeve eliminates this mechanically weaker zone, prevents excessive flexing of the bare copper zone, and thereby minimizes the chances of strand failure. The sleeve also provides electrical insulation over the bare copper strands and the barrel of the socket. This is useful in preventing accidental contact between adjacent sockets in closely spaced connections.

The case of the wire and socket without the heat shrinkable sleeves is similar to that of a solid rod shape as shown in Fig. 9.3(a). If such an arrangement vibrates, the rod is likely to fail at the reduced section at x, whereas a rod of the same section reduced uniformly over its entire length as shown in Fig.9.3(b) will not fail under a similar vibration, because the deflection is well distributed over the entire length and not concentrated in a small zone as at x in Fig. 9.3(a).

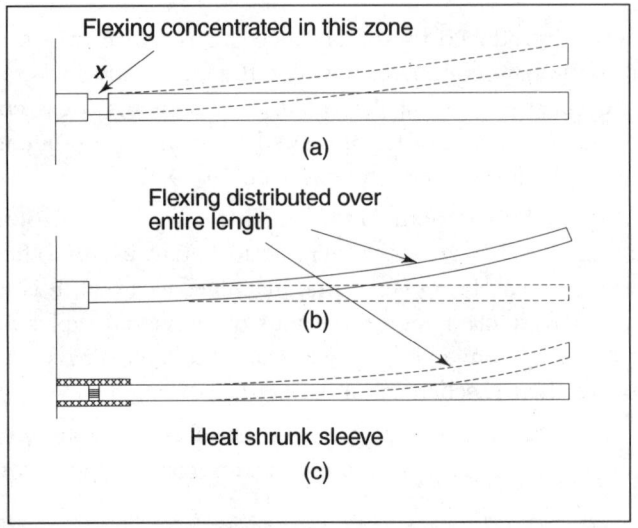

Fig. 9.3 (a) Rod with reduced section in groove (b) rod with reduced section over full length (c) wire with heat shrunk sleeve

9.4 STRAND FRACTURES DUE TO SHARP EDGES ON SOCKET BARREL

Fractures of wire strands usually occur either at the point where they emerge from the barrel of the crimped socket as shown in Fig. 9.4(a), or at the point of minimum cross-section in the crimp zone as shown in Fig. 9.5. Such fractures are more likely to occur when there is continuous relative vibratory motion between the socket and the cable cleat as shown in Fig. 8.4(b).

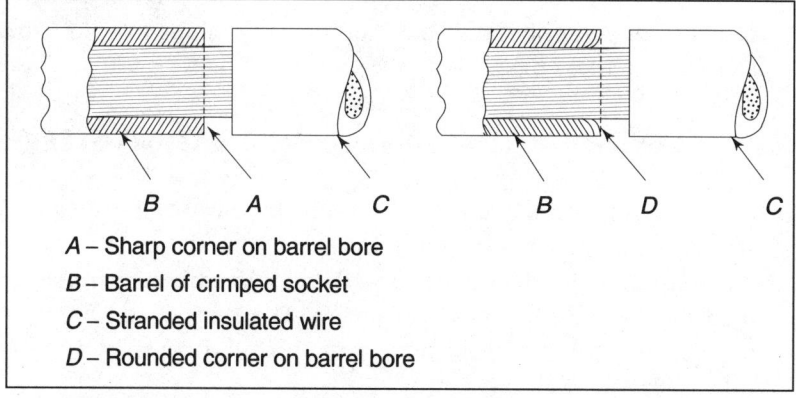

A – Sharp corner on barrel bore
B – Barrel of crimped socket
C – Stranded insulated wire
D – Rounded corner on barrel bore

Fig. 9.4 (a) Fracture of strands at corner of socket-barrel (b) Rounding of inner corner of socket-barrel

The inner edges of the barrel of the socket should be rounded off as shown at *D* in Fig. 9.4(b) to prevent any cutting action on the strands of the wire.

If a solid copper wire is bent and the radius of curvature of the bend is less than three times the diameter of the rod, there is a possibility of microcracks developing in the bent portion. The microcracks are caused by the stretching of the outer layers of the copper wires. If there is vibration, these cracks may grow and eventually cause fractures. Such problems do not usually occur on stranded cables because

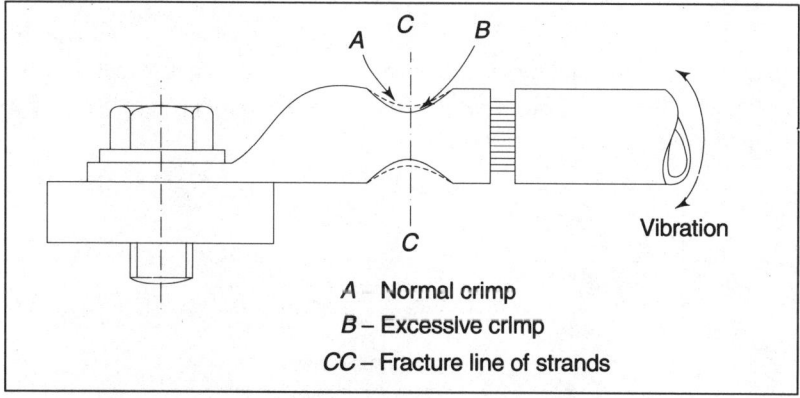

A   Normal crimp
B – Excessive crimp
CC – Fracture line of strands

Fig. 9.5 Fracture of strands through excessive crimp

the diameter of the individual strands is usually very small compared with the radius of curvature of the whole stranded cable; but bending of strands around sharp edges can produce microcracks in the outer layers of the wire, because the ratio of bend radius to wire diameter can be much less than three.

9.5 STRAND FRACTURES INSIDE INSULATING SLEEVES

Sometimes, strand fractures occur *inside* their insulating sheaths. The strands are cut neatly across as if by a shear, but there is no damage to the insulating sheath. In fact, it is difficult to locate such defects, as there are no visible signs of damage. The only observation is that there is an open circuit in one or more of the cores in a multicore cable (see Fig. 9.6).

These strand fractures are caused by fluctuating tensile stresses and metal fatigue. The fluctuating tensile stresses are produced by constant flexing of the multicore cable as in an inter-vehicle coupler. The cores on the outer radius of the bent portion of the multicore coupler get stretched. The tensile stresses produced in the strands keep fluctuating as the radius of curvature of the whole cable keeps changing (see Fig. 9.7).

Fluctuating tensile stresses which exceed the endurance limit lead to fatigue failure. If the excess over the endurance limit is high, failure will occur earlier. Calculations show that fluctuating tensile stresses produced by the flexing of multicore cables can be very much higher than the endurance limit of copper if the cores are held very firmly in the outer sheaths. If they can slide a little, the stresses are greatly reduced.

In order to permit the cores to take up positions in which the tensile stresses in the outer cores are minimum, the clamps in the coupler heads should be tightened after the multicore cables are fixed in their final positions.

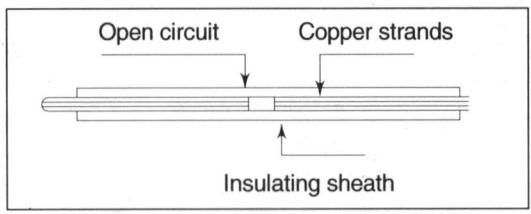

Fig. 9.6 — Open circuit inside insulation due to strand fracture

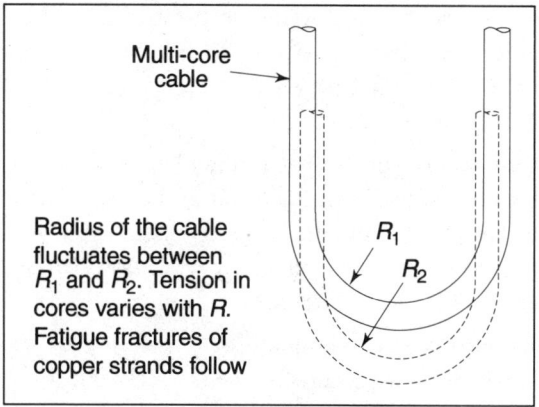

Fig. 9.7 Stretching and fracture of outer cores in multicore cables

9.6 DO'S AND DON'TS FOR PREVENTING STRAND FRACTURES

The Do's and Don'ts for preventing fractures of wire strands are:
- Fix the wire or cable directly on the vibrating machine by means of a suitable cleat.
- In case of small control cables, provide a heat shrinkable sleeve to grip the end of the socket and the wire insulation.
- Round off the inner corner of the barrel of the socket to prevent it from cutting into the strands.
- Check the crimpimg tool to ascertain whether the crimp is excessive and whether the strands are getting damaged during crimping.
- Ensure that multicore cables are clamped in their coupler heads only after they are bent to their final shape.
- Ensure that the cores in the multicore cables which are subject to constant flexing are free to slide in their sheaths.

9.7 CONCLUSION

It is possible to eliminate all possibility of wire strand failures by taking a few precautions during installation of electrical connectors, particularly at locations where there is continuous vibration.

Defects such as sharp edges on the barrel of the socket, excessive depth of crimp, absence of heat-shrinkable sleeves over the socket/cable junction, and improper location of the cleats should be considered as seed-defects.

Seed-defects do not lead to immediate failures of strands or sockets. Fractures may not occur for months or even years. Some joints may not fail at all. Nevertheless, any installation which has such defects is vulnerable if it is subject to vibration. Failures can take place without warning, long after installation.

The only sure way to eliminate all possibility of strand failures is to ensure that these seed-defects are not present.

Chapter 10

Failure of Insulation on Connector Cables

In this chapter, we will discuss:

- The modes, mechanisms and causes of failures of insulation on the cables of electrical connectors.

- The causes of insulation failure due to:
  - mechanical damage
  - electrical damage
  - thermal damage
  - chemical damage

- The factors or conditions which contribute to the initiation or acceleration of damage, for each of the four damage modes mentioned above.

- Remedial measures for the prevention or retardation of damage to cable insulation.

10.1 INTRODUCTION

Electrical connectors generally consist of pieces of insulated cables terminating in crimped sockets. There are many types of cable insulations, such as polychloroprene, butyl rubber, and silicone rubber, but most of the cable failures are generally not dependent on the type of insulation. Failures of insulated cables, unlike those of the crimped sockets, are due to external defects or deficiencies. This difference arises out of the fact that whereas cables are manufactured by automatic machines and subjected to many stages of inspection and testing, crimped sockets are installed manually, sometimes in cramped and inaccessible locations.

As we have seen in the foregoing chapters, electrical connector failures can occur due to a variety of reasons or seed-defects at the terminations, i.e., at the crimped sockets or terminals. Cable insulation failures too, are due to a number of seed-defects introduced during the installation of the cables.

If we consider the failure rates corresponding to each individual cause of failure, we find that they seem to be very small; but none of these can be ignored because the overall failure rate is the sum of a large number of small failure rates; and this overall failure rate is unacceptably high. The only way of reducing the overall failure rate is to try to reduce to zero, the failure rate of each individual cause by taking action appropriate to that cause. There is no single solution which can take care of all types and causes of failures.

Cable insulation failures due to gross design defects are not being considered here. It is assumed that cables of the required current rating and the required voltage rating are installed. This is a reasonable assumption because these cable ratings are well established and there is little possibility of error in this regard. Even if, occasionally, incorrect types of cables are selected, high failure rates in the initial stages will expose the errors and lead to corrective action.

In case of doubt, the current ratings of cables can be determined very easily from the copper section and reference to electrical handbooks. If the voltage ratings are not marked clearly, the breakdown voltage measured according to a test method given in the relevant standard will give the required information.

Most of the insulation failures on cables which normally take place in practice, are those due to degradation processes which can be classified into four groups as follows:

- *Mechanical damage.* This is a very common type of degradation process. The insulation is damaged due to chafing or rubbing of the insulation against some structural member or against another cable.
- *Thermal damage.* Thermal damage to insulation is another usual type of degradation encountered in practice. All insulating materials commonly used in cables have their own limiting temperatures which are usually in the range 90–130°C. Excessive temperature leads to premature ageing and failure.

Failure of Insulation on Connector Cables

- *Chemical damage.* The most common agent which damages the insulation on cables is ordinary lubricating oil. This substance is often present in and around machinery.
- *Electrical damage.* Excessive voltage, either continuous or transient, can damage the insulation to such an extent that it will fail eventually even at normal operating voltages.

10.2 INSULATION FAILURES DUE TO MECHANICAL DAMAGE

The problem of insulation failures due to mechanical damage is experienced mainly on equipment subject to continuous vibration, and sometimes even on stationary equipment due to thermal expansion and contraction of cables. However, these failures are due to totally avoidable seed-defects shown in Fig. 10.1.

Consider the arrangement shown in Fig. 10.1. Here, the machine and its terminal board vibrate continuously as the machine is mounted on resilient pads or springs. In such a case, there is a possibility of cable insulation getting cut at point *P* where there is contact between the cable insulation and a metallic component *D*. Even where the machine is rigidly fixed to the base as shown in Fig. 10.2, a similar failure may occur if the unsupported portion of the cable is long and subject to vibration and flexing or due to vibration of the cleat *F*.

If the connector cable length is small, it is self-supporting between the terminal sockets which are held firmly by the screws or nuts/bolts.

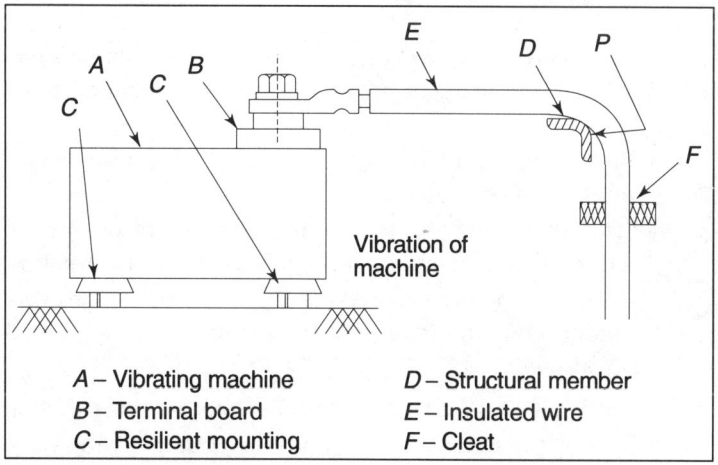

Fig. 10.1 | Insulation failures due to abrasion of cable insulation (due to machine vibration)

There is no possibility of insulation failures unless the cable rubs or bears hard against some other components. While vibration is often the proximate cause of failure, the seed-defect which is the true cause of the insulation failure is the direct contact at *P* between the cable insulation and some other metallic component as shown in Fig. 10.1 and 10.2. If such contact is unavoidable due to space or other restrictions, or if the cable length is such as to need intermediate support, the cable should be supported in a properly designed cleat.

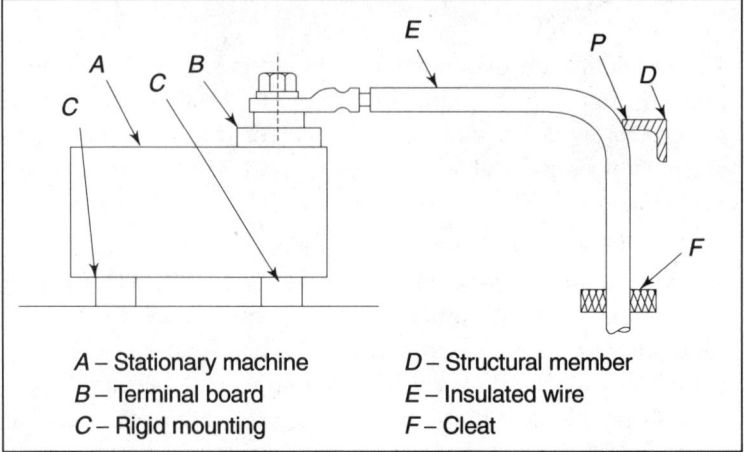

A – Stationary machine
B – Terminal board
C – Rigid mounting
D – Structural member
E – Insulated wire
F – Cleat

Fig. 10.2 | Insulation failures due to abrasion of cable insulation (due to cable vibration)

Vibration, usually the proximate cause of failure, can be of different types as follows:
- One end of the electrical connector may vibrate due the vibration of the machine to which that end is connected. The other end may be free from vibration.
- If the unsupported length of the cable is excessive, the middle portion of the cable may start vibrating.
- The alternating elongation and contraction of the cable due to changes in its temperatures is a very slow process, but the forces produced by thermal expansion or contraction are very high and under certain circumstances, failure can be traced to such processes.

If the insulation of a vibrating wire rubs or bears hard against any metallic component, the insulation will wear out physically and then fail electrically when the insulation thickness gets reduced to a level at which it punctures at normal operating voltages.

Failure of Insulation on Connector Cables

A very effective and simple way of preventing insulation failures due to vibration and wear is to clamp the insulated wire(s) at suitable intervals in such a way as to ensure that there is no contact between the cable insulation and any structural member or other component at any point. It should also be ensured that if a cable must absorb some vibration, as in the case of a connection from a vibrating machine, the cleats should be so placed as to allow the vibration to take place on the cable between two cleats. Further, the portion of the cable which is allowed to flex should not come into contact with any component. These simple design rules are shown in Fig. 10.3.

If a cable is passed through a hole in a steel plate or sheet without any protection, as shown in Fig. 10.4, it constitutes a seed-defect which can grow into a defect and a failure. The insulation on the cable will get damaged in course of time due to vibration or thermal expansion and contraction of the cable while pressing on the steel plate.

Failures of this type should be prevented by fixing a grommet (grooved sleeve) or cleat as shown in Fig. 10.5. These arrangements will prevent mechanical damage and also reduce eddy current heating of the plate due to the larger opening.

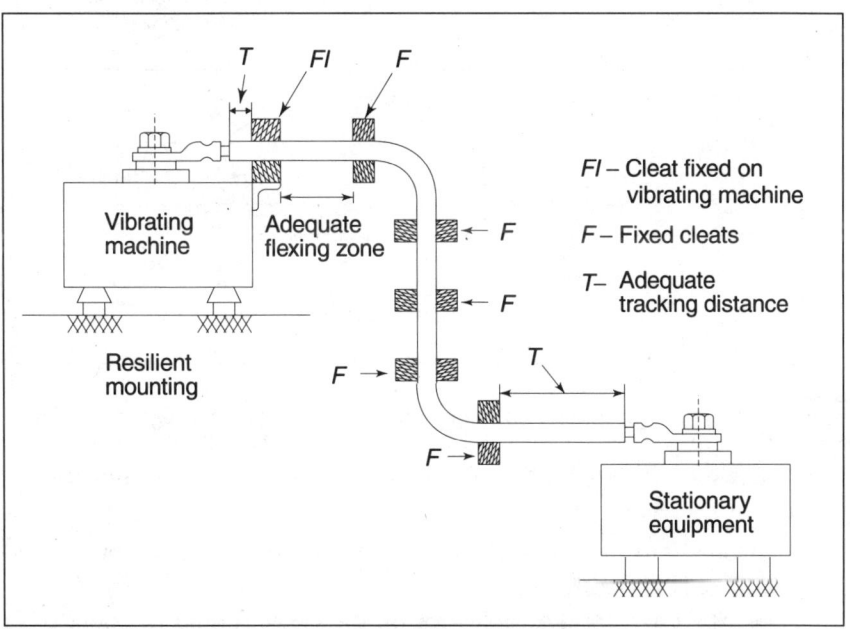

Fig. 10.3 Correct method of clamping cable to prevent failures

The measures listed above for preventing insulation failures may appear to be so simple and practical that any special mention may seem to be unnecessary; but the fact is that such seed-defects do get introduced in many installations. In most cases, failures do not take place for months or even years, but sooner or later, and sometimes

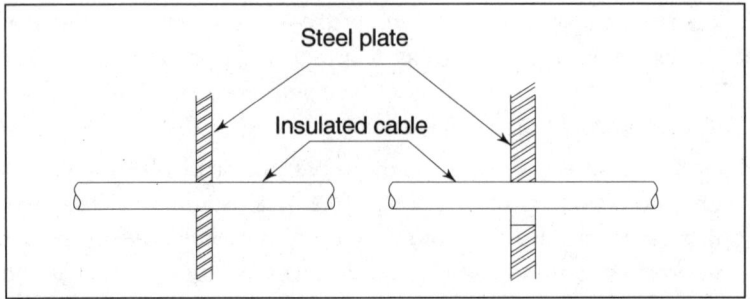

Fig. 10.4 Insulation failure on cable passing through steel plate

unexpectedly, failures do take place. The only way to prevent even these rare failures is to train the workers suitably in this regard and to make special checks to see that throughout the cable runs, there is not

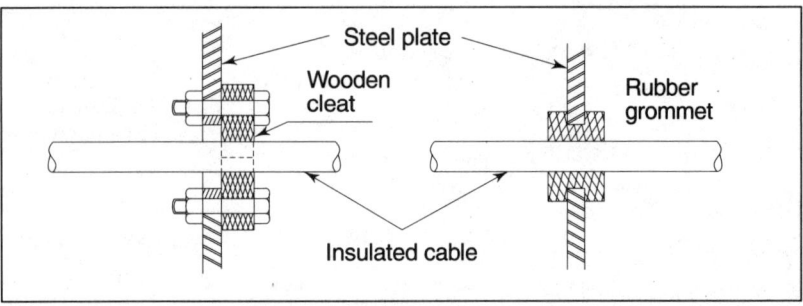

Fig. 10.5 Preventing insulation failures on cables passing through steel plates

a single location where such seed-defects are allowed to remain and go into service.

It only remains now to discuss a few apparently minor but important details about the design of the cleats.
- The edges of the cable seat on the cleats should be rounded to remove any sharp corners which could cut into the insulation (see Fig. 10.6).

- Where the cable connection is with a vibrating machine, elastomeric liners should be inserted between the cable and the cleat.
- The radius of the groove in the cleat should be equal to the cable radius and the depth of the groove should be less than the radius, so that the two halves of the cleat can hold the cable(s) firmly in their grip. The cables must not be allowed to remain loose in the grooves (see Fig. 10.7).

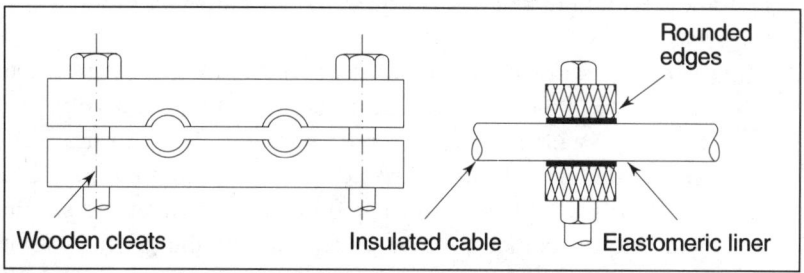

Fig. 10.6 Rounding of corners on cable cleats to prevent insulation damage

10.3 INSULATION FAILURES DUE TO THERMAL DAMAGE

It is a well-known property of solid insulating material that their life is halved for every 8°C rise in temperature above their index temperature corresponding to life of 20,000 hours.

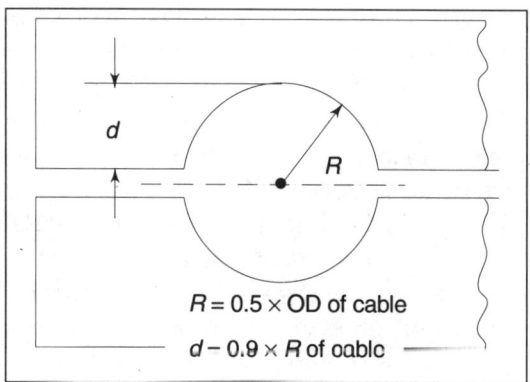

$R = 0.5 \times$ OD of cable
$d - 0.9 \times R$ of cable

Fig. 10.7 Cleat dimensions to ensure firm gripping of cable insulation

Thus, for example, an excess of 80°C over the index temperature would reduce the life to:

$20000/2^{10}$ hours = about 19 hours

It may be appreciated that failures of insulation do not take place immediately after any increase in the temperature. The degradation may take months or even years before the condition reaches a stage where it fails at normal operating voltage. Cable insulation can be degraded thermally due to overheating, in many different ways as follows:

- Normally, a power supply in accordance with the Indian Electricity Act and Rules is supposed to be maintained at a voltage and frequency which is within certain specified limits. In the prevailing conditions in India, this is not always ensured by the Electricity Boards. A voltage and/or frequency lower than the specified limit is quite common. Voltages higher than the permissible limit are also experienced from time to time. Frequency higher than the permissible limit is rare. When the voltage is 'low' (lower than the permissible limit), some equipment like constant speed motors draw currents which are higher than the rated values. Similarly, if the voltage is 'high' and/or the frequency is 'low', equipment like transformers draw higher magnetising currents. If such contingencies are not provided for in the design of cables and terminals, they can get overheated in service under abnormal supply conditions.
- Use of electronic equipment for regulation and controlled variation of electrical supply is becoming increasingly common. From small items like ceiling fans, fluorescent fixtures, television sets, and personal computers to large high power units for welding machines, machine tools, furnaces, and variable speed motor drives, such electronic devices are used in increasing numbers. While they control or regulate the 'output' beautifully, these devices play havoc with the 'input' system by introducing harmonics in the input current, the first direct effect of which is to increase the heating effect on cables and terminals. If the designer fails to take these factors into account, overheating of cables and terminals can take place.
- If the cable cross section is too small for the current it is required to carry, it will obviously get overheated; but this is due to a gross design defect.

Failure of Insulation on Connector Cables

- If a cable is installed too close to a device (e.g., a resistor or a steam pipe) which gets very hot, the cable insulation may get locally overheated and eventually fail at normal voltage, due to degradation of the insulation. Another possible cause of local overheating and consequent failure of insulation, is the heating of PVC insulation on cables beyond 70°C due to conduction of heat from terminals which are designed to operate at 90°C.
- If a cable carrying a very high alternating current is installed very close to a thick steel member, it can get heated to a high temperature as a result of eddy currents in steel. The insulation will be damaged and will fail at normal operating voltage.
- If a cable or group of cables originally designed to operate in free air gets covered in dust or similar substance which prevents free dissipation of heat, the cable will get overheated and the insulation will get damaged. There will be insulation failure eventually at normal operating voltage.

The six possible causes of local overheating and damage to insulation mentioned above are not hypothetical cases. They are taken from actual case histories. There are many other possible ways in which cables can get overheated locally. When investigating cable insulation failures, possibilities of local overheating should be looked into.

10.4 INSULATION FAILURES DUE TO CHEMICAL DAMAGE

Insulation failures can also occur due to local degradation of the insulation due to the effect of oil on the insulation. Lubricating oils which may come into contact with cables due to leakage or spillage soften the insulation of many types of cables. Oil-resisting protective sheaths are often provided on cables but even these may be damaged although after a longer period of time.

Cable insulation may be damaged by other chemicals too. This possibility should therefore be kept in view, particularly when investigating cable faults in chemical industries.

10.5 INSULATION FAILURES DUE TO ELECTRICAL DAMAGE

If the voltage rating of a cable is much lower than the normal operating voltage, the cable will obviously fail sooner or later. However, this type of gross and obvious design error is unlikely.

Unexpected rise in voltage must, however, be considered when investigating cable insulation failures, specially if there are no signs of damage due to any of the three other causes mentioned above. Some of the possible ways in which this can occur, taken again from actual case histories, are as follows:
- When there are several independent voltage systems operating side by side, cables in the lower voltage system can be affected by contact with the higher voltage system due to some failure of some other component. A common example of this is the power and control systems in electric locomotives.
- Transient high voltages such as those due to lightning, switching surges and electromagnetic induction from nearby voltage systems, are some of the possible ways in which insulation can be damaged electrically.

10.6 REMEDIAL MEASURES AGAINST DAMAGE TO INSULATION

The remedial or preventive measures against mechanical damage have already been referred to in Section 10.2. As regards the other three classes of damage to insulation, there is no simple solution. The measures to be taken will depend on the actual cause of damage and the local conditions; once the cause is established the remedy becomes obvious at once. The factor or agent which causes the damage to insulation must be eliminated, or at least kept away from the insulation.

10.7 DO'S AND DON'TS FOR PREVENTING CABLE INSULATION FAILURES

- Ensure that there is no direct contact between the cable and any metallic component by providing, if necessary, suitable cable support cleats.
- Ensure that the cleat grooves are dimensioned in such a way as to grip the cable firmly, while at the same time, there are no spaces left to permit cable vibration within the cleat.
- Ensure that the edges of the cleats are rounded so as to prevent cutting into the cable insulation. Provide elastomeric liners if necessary.

Failure of Insulation on Connector Cables 169

- Ensure that there is no localized overheating of cables due to external heat sources such as eddy current heating of steel plates in the vicinity, resistors, steam pipes, etc.
- Ensure that the dissipation of heat from cables is not impeded by the accumulation of dust, ash, etc.
- Ensure that lubricating oil or other harmful chemicals are not allowed to come into contact with the cable insulation.

10.8 CONCLUSION

Since insulated cables are manufactured by automatic machines and since the process of manufacture includes many stages of inspection and testing, defective insulation is rarely the cause of cable insulation failures.

Cable insulation failures are generally due to seed-defects introduced during the manual installation of the cables.

If the six seed-defects enumerated above are eliminated from the installation, cable insulation can be made totally free from failures. On the other hand, if such seed-defects are allowed to remain in service, the cables will always be vulnerable and they may fail without warning, months or even years later.

Chapter 11

Failures of Terminal Boards

In this chapter, we will discuss:

🔌 The common features of terminal boards of different designs insofar as they are relevant to failure mechanisms.

🔌 Failure modes of terminal boards, which lead to electrical fires and failures.

🔌 Common defects in design, manufacture and installation of terminal boards which eventually lead to failures.

🔌 The mechanisms of failures arising out of the common defects.

🔌 Measures or precautions to be taken during design, manufacture and installation, to prevent terminal board failures.

🔌 Special design features of totally reliable terminal boards.

🔌 11.1 INTRODUCTION

Although no high technology is involved in the design, manufacture and maintenance of terminal boards, it is a fact that many electrical failures and even major electrical fires start from seed-defects in terminal boards. Unfortunately, these seed-defects are often not perceived in their potentially dangerous form.

All electrical equipment are usually provided with terminal boards where the internal and external cables are connected together, usually with threaded fasteners. Terminal boards facilitate replacement of complete equipment and also troubleshooting.

Failures of Terminal Boards

A terminal board is basically an insulating board with a number of bolts/nuts, screws, or studs fixed in line in an accessible location. Figure 11.1, shows three terminals. However, terminal boards may actually have many more terminals.

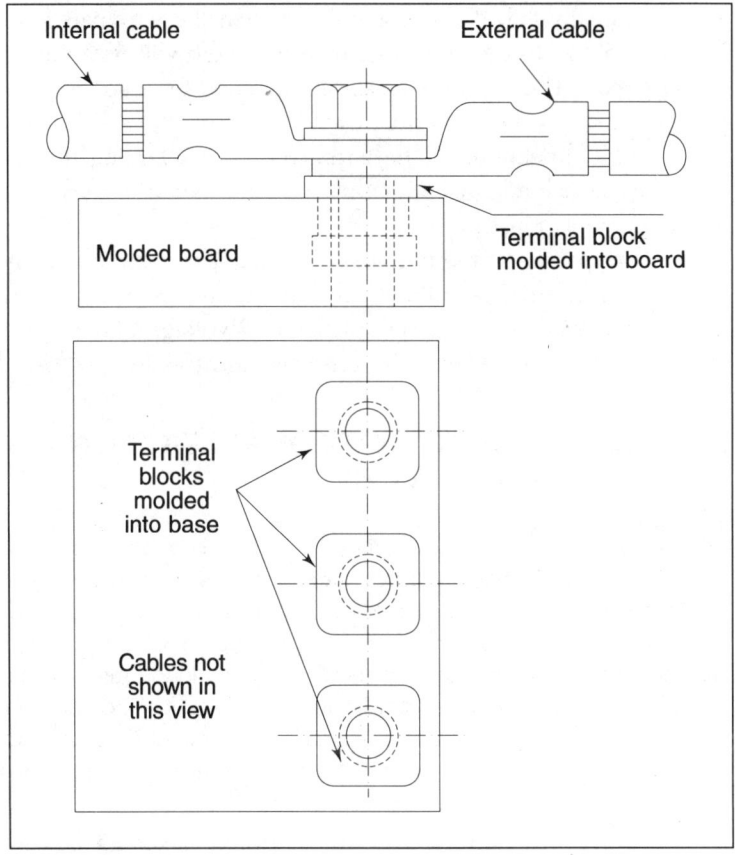

Fig. 11.1 Typical terminal board of good design

Terminal board failures are usually due to one or more of the following causes:
- Inadequate size of the terminals.
- Inadequate tightening of the fastener.
- Shrinkage of the terminal board laminate.
- Tracking between terminals.
- Metal creep.

11.2 FAILURE DUE TO INADEQUATE SIZE OF TERMINALS

Inadequate size of terminals is a major design defect. If the terminals are too small for the current to be carried, they will get overheated and will eventually fail. If the distance between the terminals is too small for the voltage to be withstood, then too there will be a failure due to short circuit. These problems can be avoided by following the design and test criteria mentioned below:
- The temperature of the terminals, when carrying the maximum possible load current, should be less than 90°C at the maximum ambient temperature.
- The spacing of the terminals and the size of the board should be such that the terminals can withstand a 60 Hz test voltage of three times the maximum operating voltage plus 1 kV, for one minute applied between terminals and to earth.

11.3 FAILURE DUE TO INADEQUATE TIGHTENING OF THREADED FASTENERS

Terminal board failures are generally due to processes or mechanisms of failures which are similar to those encountered in the overheating of crimped sockets at the ends of electrical connectors due to insufficient crimping. These have been discussed in Chapter 6. It may be recalled that the root cause of the failure was inadequate contact force between the conductor strands and the socket. The most common cause of terminal board failures is also inadequate contact force between the components carrying current. One reason for this is inadequate tightening of the fasteners.

Inadequate tightening of terminal board fasteners is also a rare occurance because electricians learn, in due course, the importance of tightening fully all the nuts and screws used for making electrical connections. Errors are usually committed by those without adequate experience and by those who are not provided with appropriate tools. The only way to prevent such errors is to provide practical training and appropriate tools to all staff before they are allowed to work on electrical equipment.

The importance of systematic training of all staff on such simple or apparently trival matters cannot be overemphasized. Many of the important requirements of ordinary items of work are often not self-evident and newcomers make mistakes simply because no one teaches them the correct procedures. They would learn the procedures eventually but at the cost of failures in service. Since there would be some new staff all the time in large organizations, it is necessary to ensure that no one is allowed to work without prior training on such matters.

A special case in which inadequate tightening of fasteners is observed is where the screw or bolt is made of copper or brass. These metals have a relatively low yield point and any effort to tighten the fasteners fully leads to yielding or plastic flow at points of high stress, with the result that the contact force gets reduced during the tightening process itself. In extreme cases, the screw or bolt may even fracture. Fear of this problem may inhibit proper tightening by the workmen. It is best, therefore, to avoid the use of these metals for fasteners used in bolted connections. Sometimes, atmospheric corrosion may come in the way of using mild steel hardware. In such cases, either stainless steel or galvanised steel may be used. Copper and brass must be avoided.

Another possibility of inadequate tightening arises where blind tapped holes (closed at bottom) are used for bolted connections as shown in Fig. 11.2.

The normal design will always provide adequate clearance at the bottom of the tapped hole, but this may be lost either because the

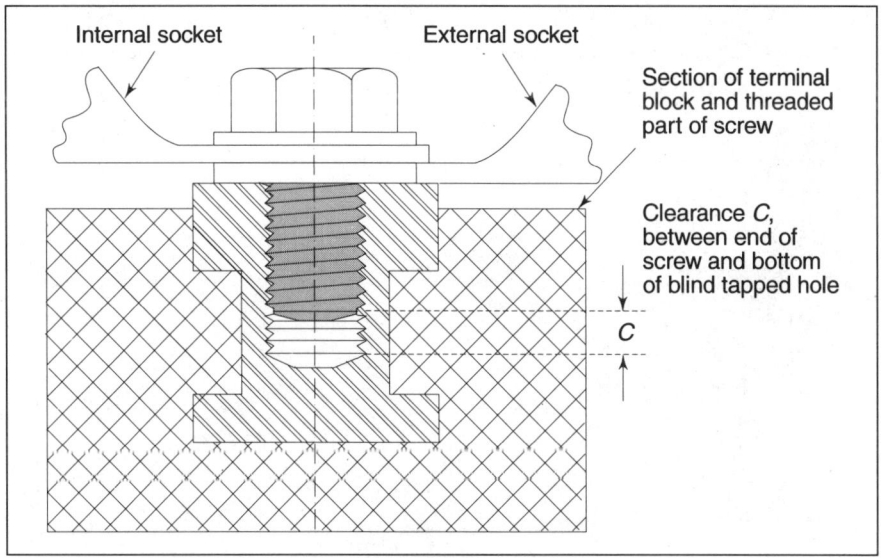

Fig. 11.2 Essential clearance in blind tapped holes

depth of tapping is less than the specified value, or because the length of the screw is more than the specified value, or because specified washers have not been inserted during assembly. Yet another possibility is that foreign matter, machining swarf etc. may have collected in the blind holes. If the problem is very severe, it may get detected during assembly; but if the interference is marginal, there is every possibility that the screw may appear to be tightened fully, whereas actually this may be against the bottom of the tapped hole. In such a case, the sockets to be connected together would not be subjected to adequate contact force. There are many cases of failures of this type due to this apparently trivial defect. Therefore, blind tapped holes with screw should be avoided for terminals. If their use is unavoidably, it should be ensured that there is a clearance of at least 2 mm at the bottom of the tapped hole when the screw is fully tightened.

Omission of or defects in spring washers in the force circuit of the fastener can also lead to loss of tightness of the fasteners and consequent failures, where the equipment is subjected to continuous vibration.

Yet another possibility of inadequate tightening is sometimes seen in the assembly of parallel clamps on overhead lines. Here the contact force is inadequate because of lack of adequate reaction to the tightening torque. Many cases of parallel clamp failures can be traced to this type of defect in installation. It is possible to check this point very easily as follows:

- First tighten the nut using only one wrench.
- If now a second wrench is held on the head of the bolt also, as shown in Fig. 11.3, it will be seen that the nut can be tightened

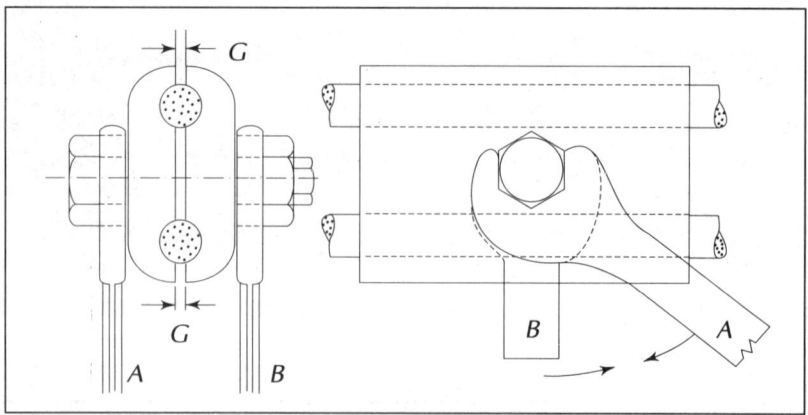

Fig. 11.3 | Use of two wrenches for tightening bolts/nuts on an overhead line

further by a significant angle showing that the original tightening was not complete.

Full tightening of the fastener can be ensured where the fastener is mounted on an unstable platform as in the case of an overhead line, by using a second wrench on the head of the bolt to provide the necessary reaction to the tightening torque. Special mention of this apparently trivial point has become necessary because the number of failures in service due to its neglect is not small. It becomes doubly important when one is aiming at zero failure performance. Gap G is essential for reliability.

11.4 FAILURES DUE TO SHRINKAGE OF TERMINAL BOARDS

The design of a typical terminal in a terminal board in an electrical equipment in shown in Fig. 11.4. A synthetic resin-bonded paper or cloth laminate supports a number of terminals side by side, of which only one is shown in the figure. This design works well for a few months, but it is almost certain to fail sooner or later, particularly if the terminals carry heavy currents. The exact nature of the seed-defect which eventually leads to such failures will be evident from the failure mechanism, which is generally as follows: the laminated insulating board which supports the terminals is under a high compressive stress produced by tightening nut D on bolt E. This tightening is essential for ensuring a low contact resistance between the internal socket A and the external socket B, and thus preventing overheating of the terminal.

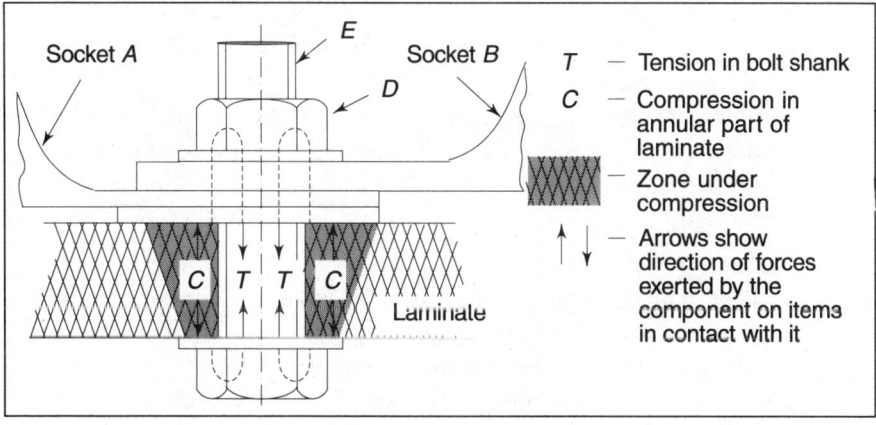

Fig. 11.4 Force circuit in a terminal board of defective design

All non-metallic materials such as insulating laminates, shrink gradually under the influence of pressure. The shrinkage is higher in magnitude than the elastic strain, or elongation in the bolt shank in between the nut D and the head of bolt A.

Let us consider an example. Assume that the relevant design details are as follows:

Thickness of the laminate board 25 mm

Length of the bolt shank 40 mm

Young's Modulus of the bolt steel 18000 kg/mm^2

Tensile strength of the bolt steel 40 kg/mm^2

Shrinkage of of the laminated board 0.5 percent

Normally, the nut would be tightened on the bolt until the stress in the bolt shank is about 75 percent of the tensile strength, or to 30 kg/mm^2. The elongation of the shank of the bolt would then be 40 × 30 /18000 = 0.067 mm.

On the other hand, the shrinkage of the laminated board would be 25 × 0.5 /100 = 0.125 mm.

The shrinkage of the board is thus much higher than the elongation of the bolt due to the tension in it. Thus, the effect of the shrinkage is that the tension in the bolt is relaxed or reduced. Theoretically, the tension could disappear completely, but long before that, the terminal would have got overheated and failed.

The force between the bolt head and the socket of the internal cable which had been produced by tightening the nut D is greatly reduced as a result of the shrinkage of the laminated board. An immediate consequence is an increase in the contact resistance R_c. The heat developed in the contact resistance increases proportionately as it equals $I^2 R_c$ and this causes the socket temperature to rise. The increased temperature aggravates the problem by the process of more rapid oxidation of the contact surfaces and also a more rapid shrinkage of the laminated board. A vicious cycle, as shown in Fig. 11.5, is created, and as a consequence, the temperature of the terminal keeps rising at an increasing rate.

The eventual result is usually an increase in the temperature of the terminal to a level which is high enough to cause carbonisation of the laminate, ignition of the laminate, and sometimes, a short circuit between adjacent termininals. The short circuit would actually be a blessing in disguise because it may trigger a tripping of power. If there

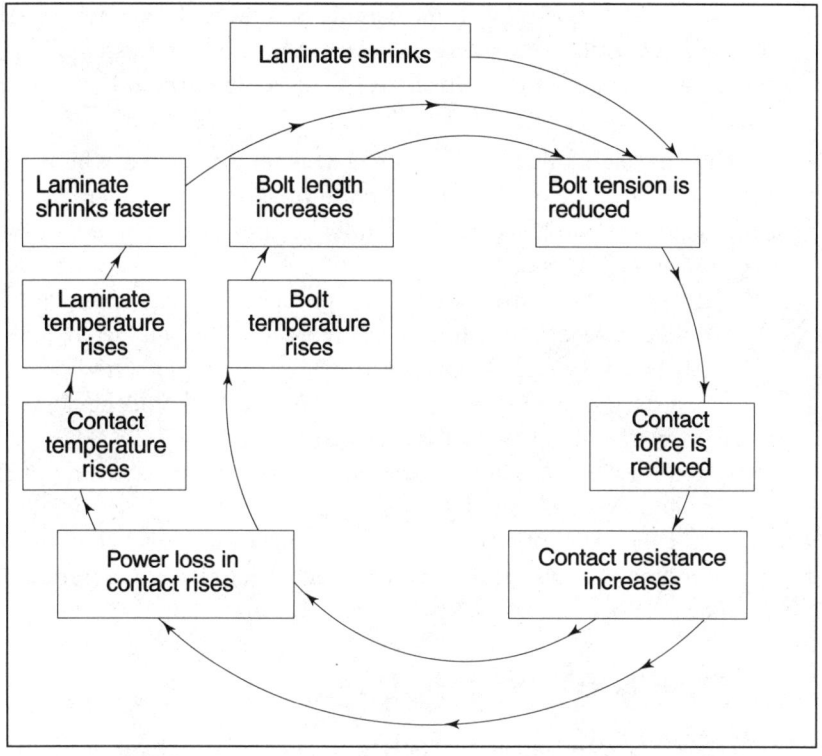

Fig. 11.5 Vicious cycle initiated by shrinkage of laminate

is no short circuit, the temperature of the bolt and nut will continue to rise until the laminate catches fire. Even if a short circuit occurs at that stage and the power trips off, it may be too late to prevent extensive damage. It may be recalled that there is no device avaialble to detect local overheating of electrical connectors which develop high contact resistance.

The mechanism of failure described above is responsible for many failures and fires. Many electrical equipment in the market continue to have these defective designs of terminals. It is possible to prevent such fires by redesigning the terminal board. Therefore, in every case of fire in the vicinity of bolted connections, the design should be carefully reviewed to see whether there is any non-metallic material in the 'force circuit' of a bolted connection. The term force-circuit here refers to the closed path in which the tensile force produced in the screw is balanced by the compressive reaction force developed in the components held together by the screw.

Figure 11.4 shows the force circuit of a bolted connection. The tensile force in the bolt shank is balanced by the compressive force in the annular zone around the hole in the terminal board and in the sockets.

It is very likely that such a defect will be discovered whenever a terminal board gets charred or when terminals melt. In that case, the only possible remedy to prevent further failures of the same type is to take the following steps:
- A cyclic check on all similar equipment should be made. All bolted connections should be tightened, the laminated boards inspected for signs of overheating and charring, charred terminal boards should be replaced; and this check should be repeated every two months until the design is modified.
- It is possible to eliminate the need for frequent tightening of the bolted connections and also to prevent such types of failures and fires described above by carrying out modifications in the terminal boards. The correct design of terminals is described in Sec. 11.5.

11.5 DESIGN FOR RELIABILITY

A terminal design which is totally free from the problem created by shrinkage of the insulating laminated board is shown in Fig. 11.6. It is

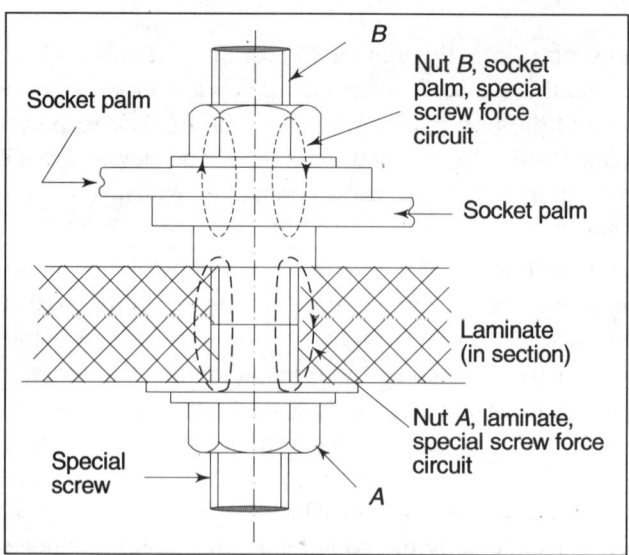

Fig 11.6 A terminal board design in which shrinkage of laminate does not affect contact force between socket

Failures of Terminal Boards

not possible to prevent shrinkage of the laminated board, but it is certainly possible to ensure that this shrinkage does not increase the contact resistance in the terminal contacts.

Here we have a special screw S with two threaded portions, of which one is used to fix the terminal screw to the laminate, and the other is used to provide a direct contact between the internal and external sockets. It will be seen that the force circuit of the latter does not include any laminate. There is total metal to metal contact and there is no possibility of any shrinkage of these metal parts at the temperatures normal to such connectors. This temperature is at most 90°C which is much below the creep temperature of either steel or copper. The screw can be made of steel as it does not have to carry any current. There is direct contact between the sockets. Any design which involves the passage of current through the bolt shank and the threads is vulnerable. Firstly, it will have to be made of copper, and secondly, any local overheating to more than 150°C may lead to creep and slackening of the contact.

The most reliable design of a terminal has the following features:
- The force circuits of the bolted connections do not include any non-metallic laminate or board.
- No current flows through the shank of the bolt or through the threads.
- The screws or bolts/nuts are made of steel. Therefore, they can be tightened to a greater extent, thereby giving a lower contact resistance.
- The temperature of the terminal does not exceed 90°C when carrying the maximum operating current.
- The screw or nut is tightened fully.
- There is adequate creepage distance between adjacent terminals.

A terminal which incorporates the above features can be expected to give zero failure performance. Of course, it is also necessary to ensure that the screws or nuts are tightened fully to begin with.

Another design which incorporates all the desirable features mentioned above is shown in Fig. 11.7. In this design, the special screw is moulded in the laminated board, thereby eliminating the need for an additional threaded shank and nut to hold the terminal securely on the terminal board.

When there is direct contact between copper surfaces which are bolted together, and if the force exerted by the bolt or screw is adequate, there should normally be no problem of bad contact even if

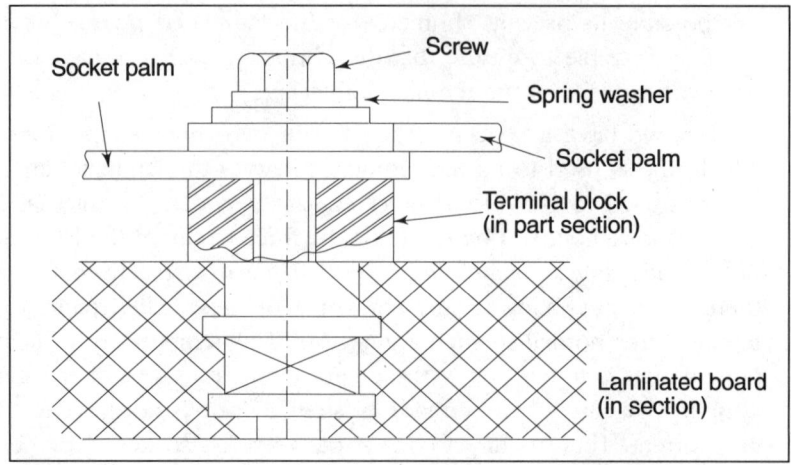

Fig. 11.7 Another design of terminal board in which shrinkage of laminate does not affect contact force

the contact surfaces are untinned. However, if there is a possibility of the temperature of the copper contacting components exceeding 90°C, oxidation of the contact faces can occur, and this can lead to a progressive increase in the temperature of the bolted joint with a consequent rise in oxidation rate and a vicious circle culminating in failure. In such cases, it is desirable to provide tinning of the contact faces. Tinned faces do not oxidise as easily as copper, and further, tin oxide has a better conductivity than copper oxide. With tinning, the temperature can be allowed to rise to 105°C. For zero failure performance, it is desirable to provide a further margin 10°C below the limits mentioned above.

In any case, whether the contact surfaces are bare copper or are tinned, it is desirable to ensure that the surfaces are cleaned well before assembly. The nuts/bolts or screws must be tightened fully. The sizes of the bolts or screws should be such as to produce a compressive stress of about 2 kg/mm^2 of the nominal area of contact. Crimped sockets must also be tinned, and preferably, tinned copper wires should be used.

11.6 IMPORTANCE OF CONTACT FORCE

The importance of contact force cannot be over-emphasised. As stated earlier, the contact resistance of any bolted connection depends mainly on the force with which the two elements of the contacts are pressed together. This force is absolutely vital for the good performance of the contact. There is an inverse relationship between the force and the contact resistance. If the force is reduced, the contact resistance increases and if the force is increased, the contact resistance is reduced up to a limit.

If the contact force is reduced for any reason, the increased contact resistance leads to an increase in the heat generated at the contact by the passage of the current. The heat generated is given by the formula' I^2R. The heat developed at the contact surface is thus directly proportional to the contact resistance R. An increase in temperature has several effects:

- If there are different metals, differences in the coefficients of thermal expansion and also in their temperatures can lead to thermal stresses. These stresses can go well beyond the yield points of the metals and cause some permanent deformation, which in turn, can lead to loosening of the bolted joint, and to consequent increase in contact resistance and eventual failure.
- If metals with low melting points (like aluminium) are used in the bolted joint, there may be metal creep at the normal operating temperature. Similarly, there may be shrinkage of non-metallic laminates under the influence of pressure and heat. All this will lead to relaxation of the force and the vicious circle which ends in a failure of the bolted joint.
- Increased temperature increases the rate of oxidation of the surfaces in contact. Copper oxide is not a good conductor of electricity. Formation and growth of oxide films at the points of contact leads to increased contact resistance and again to the same vicious circle which ends in a failure.

Thus, the starting point of contact failure is almost always a reduction in contact force. This is particularly true in the case of bolted joints. Failures due to loss of contact force may occur in several different ways. These are as follows:

- Permanent set caused by excessive thermal stress is one possible source of the problem. This can be avoided by ensuring that the initial temperature of the new joint does not exceed 90°C for bare copper and 105°C for tinned copper.
- Creep of metals of low melting point and shrinkage of non-metallic components is another usual cause. It can be avoided by using only copper for the contacts and steel for the fasteners.

- When vibration is present, the screw or nut/bolt may actually turn, thereby reducing the contact force. This has to be avoided by using locking arrangements such as spring washers, castellated nuts, etc., for ensuring a certain minimum initial tightening torque.

11.7 SPRING WASHERS

Spring washers are used to minimise the effects of the three phenomena referred to above which tend to reduce the contact force, viz. shrinkage of non-metallic components, creep of metals like aluminium, and unscrewing of the threaded fastener as a result of vibration.

When using spring washers, it has to be ensured that the sharp edges are in the right direction so that they dig into the metal and prevent loosening. Figure 11.8 shows correct and incorrect designs of spring washers. It is also necessary to ensure that the spring washers are made of the correct grade of spring steel, and that they are properly heat treated so that they do not take a permanent set and do not break up on flattening. In this connection, a reference may be made to the relevant Indian Standard on spring washers.

If spring washers are not used, or if defective spring washers are used, it does not follow that failure will immediately follow. However, when there are thousands of bolted joints in any installation and all have some defect or deficiency in the spring washer, a few are bound to fail sooner or later. Therefore, the only way to ensure zero failure performance is to see that such seed-defects are not permitted in any

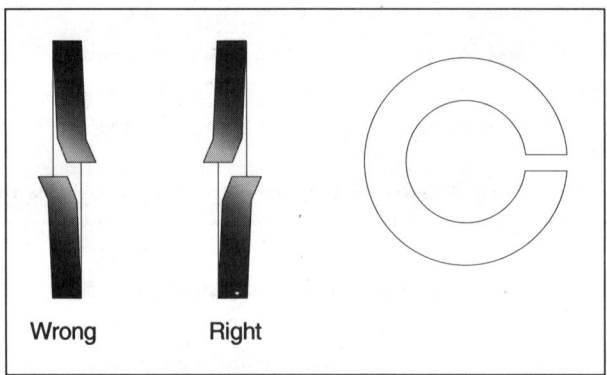

Fig. 11.8 Spring washer design detail

Failures of Terminal Boards

of the bolted connections. When there are built-in features which result in significant expansion and contraction of components in the force circuit, it is desirable to use disc spring washers (Belleville washers) to compensate for the periodic cycles of expansion and contraction.

11.8 TERMINAL BOARD FAILURES DUE TO TRACKING

The distance between terminals measured along the surface of the laminated board has to be adequate to withstand the voltage between the terminals against failure through tracking. (This phenomenon of tracking is discussed in detail in Chapter 18.) Even if the original design is satisfactory in this regard, tracking failures are likely if there are deposits of dust and moisture on the surface between the terminals. If the dust has conducting particles or soluble salts in it, the tracking process is accelerated. As far as maintenance staff are concerned, the simple requirement is to ensure that the surfaces of the terminal boards are kept clean, dry and free from any deposits.

It may, however, be noted that very often, what looks like a creepage or tracking failure is actually due to overheating and charring of the insulating material due to overheating of the terminals. In such cases, increasing the creepage distance wil not be an effective cure. It is more important to ensure that the defect which is causing the terminals to get overheated in service is eliminated.

11.9 FAILURES OF TERMINAL BOARDS DUE TO METAL CREEP

Certain metals like aluminium and lead are subject to a phenomenon called metal creep at room temperatures. Other metals with higher melting points are also similarly affected, but only at elevated temperatures. The effect of creep is similar to the shrinkage which takes place in insulating laminates under the influence of pressure, only, the creep process is much slower. Nevertheless, creep can lead to failures after years of service and, therefore, aluminium, tin, lead and such other metals of low melting point must not be used for fasteners used in terminals.

Even copper terminals are subject to metal creep if the tempeature is above 135°C. Since terminal temperature is normally limited to

90°C, creep does not lead to any problem if copper terminals are used.

11.10 DO'S AND DON'TS FOR PREVENTING TERMINAL BOARD FAILURES

- Provide direct contact between the internal and the external sockets. Tin the contacting surfaces.
- Ensure that the force circuit of the fastener which provides the contact force does not include either any non-metallic material, or any metal (e.g., aluminium) which creeps at temperatures below 90°C.
- If any terminal with shrinkable materials in the force circuits are in use, retighten the fasteners periodically. The periodicity may be determined by trials.
- Ensure that the fastener is made of steel. Use high-tensile steel fasteners if necessary.
- Do not use threaded fasteners made of non-ferrous materials.
- Use spring washers and, where necessary to compensate for shrinkage or thermal expansion, use Belleville washers under the nuts.
- Ensure that the dimensions of the terminals and other associated components are such as to ensure that the temperature rise of the terminal does not exceed 90°C when carrying the heaviest expected current.
- Ensure that the screws or nuts are tightened fully initially during installation and that the contact surfaces are neither rusted nor oxidized.
- Avoid the use of blind tapped holes and screws for terminals; but if they are unavoidable, ensure that at least 2 mm clearance is available at the bottom of the blind holes when the screws are fully tightened.
- Use two spanners in opposition to tighten bolts and nuts on unstable platforms, such as over-head lines.
- Ensure that adequate creepage or tracking distance is provided between adjacent terminals on the terminal board. The boards should be capable of withstanding a 60 Hz test voltage of $(3 \times U + 1)$ kV, where U is the maximum operating voltage in kV.

Ensure that the surfaces on the terminal board which are vulnerable to tracking failures are kept clean, dry and free from any deposits.

11.11 CONCLUSION

If the Do's and Don'ts enumerated above are followed, zero failure performance from the terminal boards can be expected.

Terminal boards will always be vulnerable to sudden failure without warning, if the following seed-defects are allowed to remain in service.
- fasteners are of inadequate size
- fasteners are not tightened fully
- operating temperature of terminal exceeds 90°C
- insulating laminate forms part of contact force circuit
- clearances between terminals and to earth are inadequate
- terminal board surfaces between terminals are not kept clean, dry and free from deposits.

Chapter 12

Failures of Welded Brazed and Soldered Joints

In this chapter, we shall discuss:

- The advantages and the limitations of welded, brazed and soldered joints in electrical conductors.

- The specific applications for which these three types of joints are most suitable.

- The modes and mechanisms of failures of these three types of joints.

- The critical details of the processes and equipment used for making these joints insofar as they are relevant to their reliability.

- Measures and precautions that should be taken to prevent the failures of welded, brazed and soldered joints in electrical conductors.

12.1 INTRODUCTION

Welded, brazed and soldered joints are generally used inside electrical equipment for permanent connections between various components and sometimes between subassemblies.

In Chapters 5 to 11, we discussed electrical connections in which the joints between different conductors are made by contact under force. As explained in detail in these chapters, the force between contacts is of the greatest importance for obtaining good and reliable electrical contact. This contact force may be developed either by threaded fasteners, or by springs or by crimping; and for reliability, the

Failures of Welded Brazed and Soldered Joints

magnitude of the force must not be allowed to fall below certain minimum levels which depend mainly on the current to be carried.

There is a very important limitation for all these types of contacts which depend on contact force for reliablility—such contacts are unsuitable for operation above 90°C if made of bare copper. The dimensions of the joint and the conductivity of the components should be such as to keep the temperature of the contacts below this temperature limit when carrying the maximum current.

With a maximum ambient temperature of about 45°C, the permissible temperature rise on copper terminals is 45°C for bare copper contact surfaces. At higher temperatures, the contacting surfaces get oxidised. Even the areas in direct contact start getting oxidised because oxygen can leak or diffuse into the microscopic spaces between the contacting components. Since copper oxide is a bad conductor of electricity, the contact resistance begins to increase and this sets up a vicious cycle of increasing temperature and increasing rate of oxidation, which culminates in overheating and failure of the joint. For this reason, bolted, crimped or pin/socket joints are not suitable for use where the joint temperature is likely to exceed 90°C.

Tinning or silver-plating of the contacting surfaces permits an increase in the joint temperature by a further 15 – 25°C respectively, but that is the limit. Where temperatures are likely to be even higher, and in locations which are not easily accessible, the most reliable joint is made either by brazing with silver-cadmium alloy, as the brazing alloy, or by butt welding, or with tungsten inert gas welding. Some examples of brazed and welded joints are:

1. Brazed joints between main field coils and interpole coils of dc traction motors.
2. Brazed joints between rotor bars and rotor end-rings of squirrel cage induction motors.
3. TIG welded joints between leads and coils of main power transformers.
4. TIG welded joints between armature conductors and commutator segments.
5. Butt welded joints between solid copper bars used for drawing into contact wire.

While brazed, welded or soldered joints are free from problems arising from oxidation and loss of contact force, they have their own problems of a different type. We shall discuss a few common defects or deficiencies in the materials and processes, which have a bearing on the reliability of welded or brazed joints.

12.2 BUTT WELDED JOINTS

There are two methods of welding that are generally in use for joining electrical conductors. These are: butt welding and tungsten inert gas (TIG) welding. In both cases, the welding is done by electric current and the factors which influence the strength and reliability of the finished joint are automatically regulated by special electronic controls in the welding machines.

Butt welding is done in a machine which consists of the following essential parts:
- An electrical power supply unit which can supply the required currents for the required durations.
- A pneumatic or hydraulic unit which can clamp the conductors to be welded, bring them together as required and press them together with the force necessary for the required duration.
- An overall control timer unit which synchronizes the operations of the two units mentioned above.

The operation of the machine is fully automatic. It is necessary only to fix firmly the conductors to be welded in the clamps provided for the purpose. The conductors should be aligned correctly and the distance between them adjusted according to the manufacturer's instructions. It is now necessary only to push the start button on the controller.

The timer control unit will now take over. The conductors will be brought together, the power supply switched on, the current regulated, the required butting force applied, power supply switched off and finally, on completion of the weld, the welded conductor released when cool enough. All this will be done in the proper sequence and at the required timings.

The factors which affect the strength and reliability of butt welded joints are:
- The magnitude and duration of the welding current.
- The magnitude and the duration of the force applied to the two components being joined together.
- The duration of the cooling period, during which the two components must be held together without any movement or vibration.
- The sequence of the above operations.

Butt welded joints are the most reliable if proper control is exercised on the magnitude and duration of the current, the magnitude and duration of the force and the duration of the cooling period.

The reliability of the welds depends on the four parameters, viz. current, force, time and sequence. These depend entirely on the reliability of the welding machine. In order to verify this, it should be made a regular practice to make test welds at the beginning of each day on small pieces of conductors and to carry out bend tests at the welds as shown in Fig. 12.1.

A butt weld is as strong as the conductor itself and the bend test can be done in the same manner as specified for the conductor. If the welded conductor passes the bend test, it is not necessary to carry out any electrical test. In fact, there would be no discernible difference in the conductivity at the weld from the rest of the conductor. In general, electrical conductors are made of annealed copper and there should be no difference in the hardness too.

If hard drawn conductors are welded, the mechanical strength at the weld would be significantly less at the weld due to annealing, as compared to the rest of the conductor. Therefore, this process is not used for joining hard-drawn copper conductors.

The factors mentioned above should be carefully controlled in accordance with the instructions given by the manufacturer of the welding equipment. However, this is not enough. For achieving the highest reliability, tests should be made with variations in these factors around the recommended values on sample test pieces to determine whether the optimum strength and reliability is being obtained with the recommended values. If the welded joint is likely to be subjected to alternating stresses, fatigue tests should be carried out on test

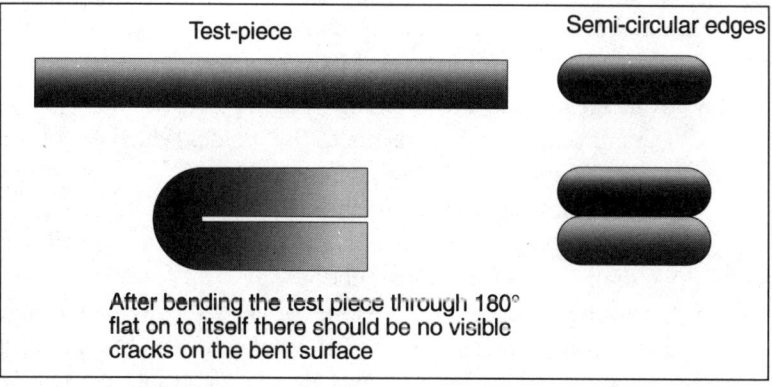

Fig. 12.1   Bend test to check butt welds

specimens to verify that the joint has a mechanical strength which is at least 95 percent of the parent material.

12.3 TIG WELDED JOINTS

Tungsten inert gas welding, or TIG welding, is also referred to as GTAW, or gas tungsten arc welding. This is basically an electric arc welding system in which the arc pool of molten metal is protected from oxidation by a shield of inert gas which is usually argon. The welds produced by this system give joints in copper conductors which are as good as the parent metal in respect of both mechanical and electrical properties.

The TIG welding equipment consists of the following essential parts as shown in Fig. 12.2.
- The welding torch consists of an electrode holder with arrangements for water cooling and for directing a stream of inert gas over the weld zone. The electrode is made of tungsten which is a metal with a melting point over 3400°C. The electrode is usually alloyed with 1 to 2 percent thorium for getting a longer life.
- Inert gas cylinder with pressure regulator.
- Electronic current regulator which regulates the welding current to any required value within close limits which are not affected by supply voltage variations and distance of electrode from the work. This feature greatly improves the quality of manual welding.
- High frequency, high voltage ignition unit. This unit provides a high voltage surge to start the arc. It is therefore not necessary to strike the arc by touching the electrode to the work and pulling out the arc. A stable arc is maintained continuously with little effort on the part of the welder.

The advantages of the TIG welding system are:
- The arc pool is clearly visible to the welder which enables better quality of the weld being made.
- There is no slag and no splatter.
- The filler metal is the same as the base metal to be welded. The welds are therefore homogeneous.
- While the welding can be done either manually or by machine, the important parameters which regulate the weld quality can be regulated within precise limits by the electronic current regulator.

Failures of Welded Brazed and Soldered Joints

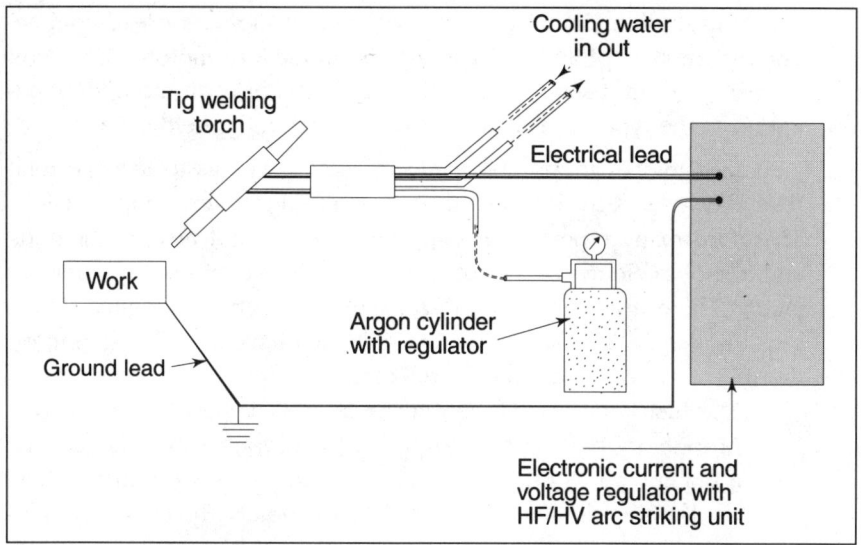

Fig. 12.2 Schematic diagram of TIG Welding Plant

The following precautions should be taken for best results when welding copper conductors:

- The welding electrode polarity should be kept negative. Positive welding electrode is not recommended for welding copper.
- The tungsten electrode must not touch the weld pool or the filler metal wire.
- Test welds should be made on sample pieces and tested by the bend test.
- The quality or purity of the inert gas should be ensured by verification at the time of purchase of the cylinders. There must be no water vapour in the gas. Hydrogen must be absent.

The verification of the suitability of materials and processes as suggested above has to be done by the designer and manufacturer of the equipment before undertaking bulk production. Maintenance engineers are also required to utilize TIG welding processes during repairs and they would be well advised to follow the detailed instructions given by the manufacturer on the basis of proven techniques.

12.4 BRAZED JOINTS

Brazed joints are commonly used in locations where the joint temperature is likely to be above 90°C, as in the case of joints between field coils and the interconnecting links of traction motors or lead connections and star point connections in induction motors. The coil temperatures are usually of the order of 130°C or even more, depending upon the type of insulating material used in the field coils.

Brazed joints can also be made in such a way as to achieve total reliability. It is necessary only to pay attention to a few simple details. There are many types of brazing alloys, fluxes and heating methods and it is possible to use any of these, provided certain precautions are taken. There are many books and manufacturers' catalogues which give the details. The important factors which have a direct bearing on the reliability of brazed joints are as follows:

- The first item here is the brazing alloy. It is best to use either a proprietory brazing alloy supplied by a reputed manufacturer or a brazing alloy which complies with the relevent industry standard. Depending upon the application details, the most appropriate type of alloy should be selected. If a proprietary brand is used, it's application engineer should be consulted. If an IS product is used, the relevant IS should be studied for guidance. Under no circumstances should a brazing alloy be selected without these preliminary investigations.
- The flux to be used is as important as the brazing alloy and this should also be selected in the same manner as the brazing alloy.
- Before undertaking bulk manufacture with any unproven materials, tests should be made to determine the mechanical strength of the brazed joints. It should be verified that the calculated or predicted strength is achieved on several samples and that there is consistency of results. In general, brazed joints are designed to develop the same strength as the parent metal. If the joint area is likely to be subjected to alternating stresses, fatigue tests should also be made to prove the design and the choice of the brazing materials.
- If the required mechanical strength is obtained, it is generally not necessary to carry out any electrical current carrying capacity tests. However, as a measure of abundant caution the voltage drop across the joint and its temperature rise may be measured while carrying the maximum current expected in the circuit.
- The mechanical strength and the electrical current carrying capacity tests referred to above are type tests and they need be done only to verify the design and the choice of the brazing materials. There are no routine tests as such which can be made on every joint. It is, however, desirable to check the visual

appearance of each joint to verify that the brazing metal has penetrated every visible junction line between the components being brazed and that there is proper wetting of the components by the brazing metal. Proper wetting is indicated by the absence of globule formation and an acute angle at the shore-line of the brazing metal as shown in Fig. 12.3(a).

- If signs of improper wetting as shown in Fig. 12.3(b) are visable it is obviously necessary to investigate the quality and quantity of the brazing materials, the temperature developed during brazing and the preliminary cleaning of the surfaces to be brazed.
- The most reliable method of heating the parts to be brazed is electrical heating as shown in Fig. 12.4.
- The heat developed can be very well regulated by controlling the magnitude and the duration of the current. The parts to be brazed can also be held together under pressure until the metal solidifies and cools sufficiently to release the tongs. These features facilitate the control of parameters which have a bearing on the reliablity of the brazed joint.
- Where the special transformer and current regulating equipment is not available, heating of the parts to be brazed by an oxy-acetylene flame is often resorted to. It is difficult to regulate the heating when this method of heating is used. There is a more serious problem that can arise when using oxy-acetylene gas flames for heating copper with dissolved oxygen content. The copper can become very brittle due to hydrogen embrittlement. Therefore, special care has to be taken, when brazing oxygen bearing copper with gas flames. The flame should be adjusted to a length of about 20 cm, and the oxygen level controlled to give

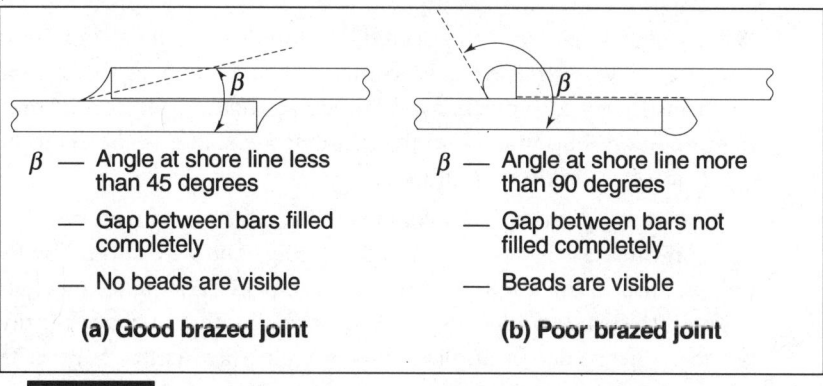

Fig. 12.3 (a) Features of a good brazed joint (b) Features of a poor brazed joint

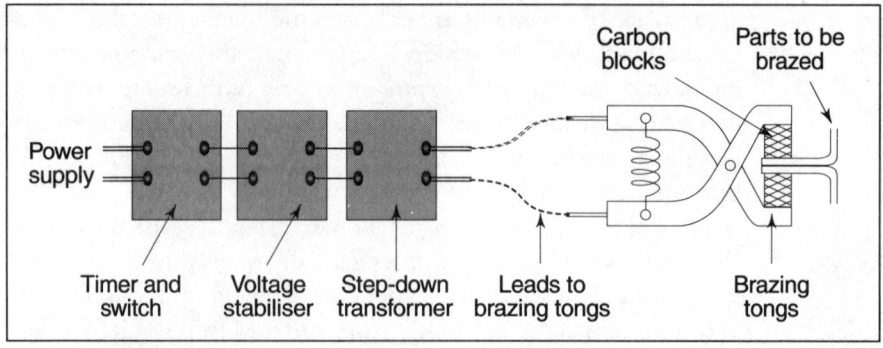

Fig. 12.4 Schematic diagram of electric brazing machine

a colorless or light blue flame. Only 6 cm at the tip of the flame should be allowed to come in contact with the copper. This will ensure that there are no reducing components in the portion of the flame which is in contact with the copper. To confirm the suitability of the heating process, a small test piece of the copper should be heated in the flame for about 10 minutes and allowed to cool. A reverse bending test should then be carried out on the annealed piece of copper in accordance with the relevant IS.

- The clearance between the parts to be brazed should be maintained within the limits specified in the drawings. Excessive clearance, as also insufficient clearance, are both to be avoided.

12.5 SOLDERED JOINTS

Where high reliability is expected, soldered joints must not be used because they are intrinsically vulnerable to certain types of failures due to the very low mechanical fatigue strength in tension of the solder and, more importantly, to the fact that the solder is subject to metal creep at operating temperatures.

If soldered joints are unavoidable, it is necessary to ensure that no mechanical stress is developed in the solder. The wire should be firmly twisted round in such a way as to ensure adequate mechanical strength to sustain whatever forces or vibration that may appear during service. The solder is applied only for electrical contact. Even then, some stress will appear in the solder due to differences in the coefficients of thermal expansion of copper and solder and this will lead to the eventual failure of the soldered joint.

A tin-lead (60–40) solder has a life of only 1000 hours at a stress as low as 0.4 kg/mm^2 at 100°C. At this stress and temperature, the life of copper is practically infinite.

For the reasons explained above, soldered joints are now rarely used in modern electrical equipment. One exception, however, is the printed circuit board (PCB). There are many soldered joints here but the construction of the PCB is such as to minimise tensile stresses in the solder. Further, most of the components are very light and where necessary, the components are clamped mechanically to the PCB. The currents are also very small and there is very little rise in temperature.

Soldering of PCBs, either manually or in a wave-solder machine, involves many practical details which are very relevant for the reliability of the finished product. A few of the more important ones are enumerated below:
- Degreasing and pre-cleaning of the surfaces to be soldered.
- Composition and purity of the solder and the flux.
- Temperature of the soldering iron or the solder bath.
- The duration of the cooling period during which the soldered joint is allowed to cool free from any vibration or stress.
- All components should be firmly fixed to the PCBs so that the soldered joints are not subjected to any mechanical stresses.

12.6 DO'S AND DON'TS FOR WELDING, BRAZING AND SOLDERING

Ensure that the butt welding, TIG welding and brazing equipment are in proper working order and that the timer/current/force/sequence control equipment in particular, are operating accurately.

For Butt Welding

- Ensure that the following parameters are adjusted correctly:
  (a) Duration and magnitude of welding current.
  (b) Duration and magnitude of butting force.
  (c) Sequence of (a) and (b) above.
- Verify the manufacturers' recommendations for the process parameters by carrying out prototype tests on samples.
- Make a daily butt welding test on a test piece and subject it to the bend test.

For TIG Welding

- Ensure that the following parameters are adjusted correctly for TIG welding.
  (a) Welding current (DC) and filler metal size.
  (b) Purity of the argon gas.
  (c) Welding electrode polarity should be negative.
- Make a daily TIG welding test on a test piece and subject it to the bend test.

For Brazing

- Ensure that brazing alloy and flux are obtained from reputed manufacturers or tested as per the relevant IS.
- Ensure that the brazing current and time are adjusted to values which have been proven by tests.
- Check the brazed joints visually to verify proper penetration of brazing metal and proper wetting of the surfaces.
- Avoid gas flame brazing as far as possible. Use electric brazing tongs.
- If gas flame heating cannot be avoided, use neutral flame and apply only the tip of the flame to work. Test for hydrogen embrittlement by bend tests on test pieces.

For Soldering

- Avoid soldered joints except in PCBs.
- If a soldered joint is unavoidable, ensure that there is no mechanical stress on the solder by twisting or wrapping the wire suitably.
- Use non-acidic, resin-based fluxes. Thoroughly degrease and clean the surfaces to be soldered.
- Regulate the temperature of the soldering iron. Check soldered joints visually to verify penetration and proper wetting of the surfaces.

12.7 CONCLUSION

Bolted connections are not suitable in locations which are not accessible for regular inspection, or where operating temperatures of the joints are higher than 90°C. Welded and brazed connections are used for making permenant connections in such locations.

Failures of Welded Brazed and Soldered Joints 197

Soldered joints have poor mechanical strength, and they should be avoided as for as possible. They are widely used, however, in printed circuit boards, on account of their compact size and suitability for mass production through the use of wave-soldering machines.

These three types of joints can give totally reliable service, but only if certain precautions are taken in their design, manufacture and installation.

Chapter 13

Metal Fatigue

In this chapter, we will discuss briefly:

- The variety of components of electrical and mechanical equipment which fail in service through a material degradation process known as metal fatigue.

- General definition and description of metal fatigue.

- Elementary theory of stress, strain and fatigue in metals.

- Introduction to the terms endurance limit, S–N curve and stress concentration factor.

- Conditions favourable to the initiation and acceleration of metal fatigue.

- Steps which should be taken to prevent the occurence of equipment failures through metal fatigue.

13.1 INTRODUCTION

It is estimated that more than 80 percent of all fractures of mechanical and electrical components are caused through a phenomenon known as 'metal fatigue'. Some of the failures which occur in modern industrial equipment as a result of metal fatigue are:
- Fractures of crimped sockets and wire strands
- Fractures of shafts and axles
- Fractures of screws and bolts
- Fractures of structural members
- Fractures of springs
- Fractures of induction motor rotor bars

- Fractures of gear teeth
- Failures of ball and roller bearings

Metal fatigue can be defined as the progressive, localized and permanent structural damage which occurs in metals when subjected to fluctuating tensile stresses and strains. In the initial stage, fatigue damage is invisible and undetectable. At the end of this stage, one or more microscopic cracks are initiated. There is a progressive growth of the crack(s) during the next stage. The cracks can be detected in this stage by ultrasonic, magnetic-particle or dye-penetrant testing. Sometimes, they are even visible. As the crack progresses, the residual intact section diminishes in area. The stress, which is equal to force/area, increases, and eventually exceeds the strength of the material. When this happens, there is a sudden fracture across the reduced section.

In the case of ferrous materials, fatigue damage takes place only when the peak stress is higher than the endurance limit of the material. In the case of non-ferrous materials, there is no such clear limit, and fatigue failures can occur even at very low stresses, but then the number of stress cycles would be very large. The endurance limit for non-ferrous metals is defined as the stress which causes failures after 500 million alternations.

Fatigue failures can be prevented by proper design and careful manufacture. Design aspects are beyond the scope of this book but some important design principles will be discussed to the extent they are relevant to manufacture and maintenance of electrical connectors. Even when the components are correctly designed, it is possible for failures to occur on account of defects or deficiencies in manufacture and maintenance.

The basic principles of metal fatigue are simple and easy to follow. It is proposed to explain these basic principles of fatigue before going into the steps to be taken for preventing failures of electrical connectors due to metal fatigue.

13.2 TENSILE STRESS

It is necessary to understand clearly the term 'tensile stress' before attempting a study of metal fatigue.

Imagine a ceiling fan suspended from the ceiling through a long steel pipe. Is it safe ? Is it possible that the pipe may break and the fan may come crashing down ? To answer these doubts, we have to make

a few simple calculations. Obviously, the answer will depend on the following factors:
- The weight of the fan.
- The dimensions of the pipe.
- The quality of the steel used for making the pipe.

Assume that the ceiling fan has a weight of 25 kg and that the pipe has a cross-sectional area of 75 mm^2 at its thinnest section, viz. the root of the threads (Fig. 13.1). We then calculate the stress on the material as 25/75 = 0.33 kgf per mm^2. Tensile stress is the force withstood by the material per unit area.

It is called tensile in this case because the material is under tension or pull. If a component is pressed by the forces acting on it, we have the material under compression. Here too, the compressive stress is calculated as the compressive force per unit area.

Let us say that the data supplied by the manufacturer of the steel pipe tells us that the breaking strength of the pipe is 35 kgf per mm^2. In other words, the pipe will break when the pull on it exceeds $35 \times A$, where A is the minimum cross-sectional area of the pipe. Since the tensile strength of the pipe (35 kgf per mm^2) is much higher than the tensile stress (0.33 kgf per mm^2), we can be certain that the pipe will not break under the weight of the fan.

The calculation given above is merely for the purpose of explaining what is meant by the term tensile stress. There are other factors and more

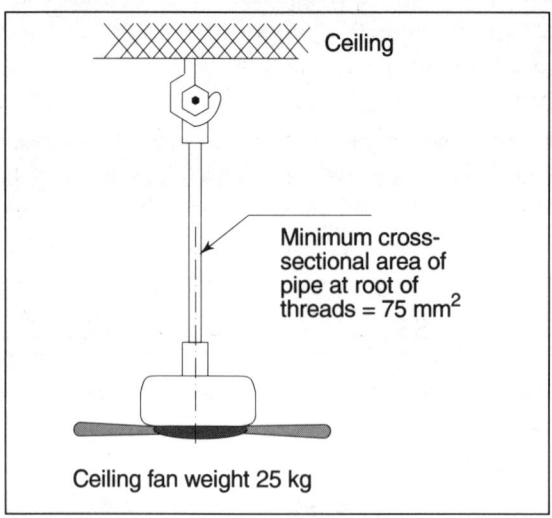

Fig. 13.1 Tensile stress in suspension rod of a ceiling fan

complicated calculations which also need to be made to answer the question about the safety of the fan suspension, but these are beyond the scope of this book. For our purpose, it is sufficient to only remember that the tensile stress on a component due to a straight pull is equal to the pull on the component divided by the cross-sectional area.

The concept of 'stress' is extremely important. It is measured in kgf per mm^2. As we have seen before, stress depends on the total load and on the cross-sectional area of the component. This stress must be much less than the 'strength' of the material, which is also measured in kgf per mm^2. In the simple example of the suspension rod of a ceiling fan, we considered a straight load along the direction of the pipe. If the direction of the load or other details of the machine are different, the calculations become more complex, but even so, we always compare the stress produced by the load with the strength of the material. The latter must always be much higher than the former. Otherwise the component may fail in service.

The simple calculation given above is for a straight pull on a rod or pipe. The calculations get more complicated if there is bending of the rod as shown in Fig. 13.2. When there is bending of a component, the stress pattern is different. There is tensile stress on one side and compressive stress on the other. At the centre, the stress is zero. If the component is vibrating or rotating, the stresses will keep reversing as shown in Fig. 13.3. It is necessary only to appreciate that when any component is subjected to external forces, various types of stresses are produced in the component; and they may vary or alternate with time.

13.3 STRESS CONCENTRATION FACTOR (SCF)

The calculation given in Sec. 13.2 gives the average stress on the material when under direct tension. There are more complicated methods

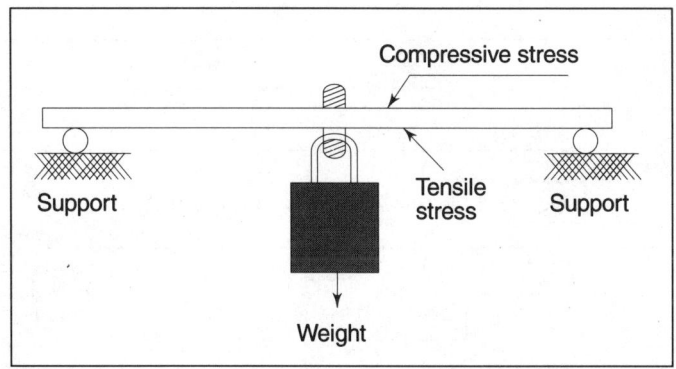

Fig. 13.2 | Tensile and compressive stresses due to bending

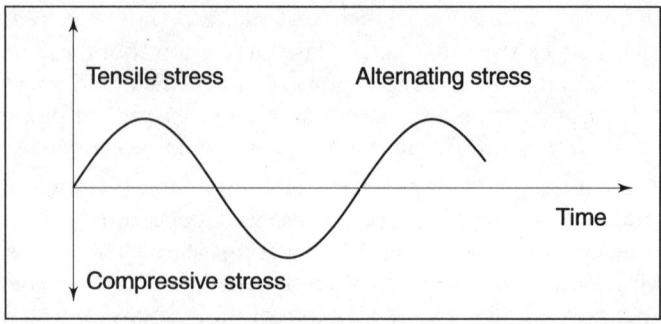

Fig. 13.3 Alternating tensile/compressive stresses due to vibration

for calculating the stresses produced by bending. The first stage of calculation gives results which are valid only if the component has a uniform cross-section as shown in Fig. 13.4. (Moreover, the endurance strength of the material as declared, is for samples which have a highly polished surface, free from any scratches, marks or corrosion.)

In actual practice, components do not usually have a uniform cross section. There are usually some changes in cross section along the length of the component. For instance, a shaft may have grooves or shoulders, or there may be holes or keyways (see Fig. 13.5).

The effect of these changes in cross section is that there is an increase in the stress by a multuplying factor known as the *stress concentration factor* (SCF), over and above the increase due to reduction in area. This factor depends on the nature of the changes in cross section. If the change in section is gradual, the factor is small; but if the change in section is sharp or sudden, the factor is large (see Fig. 13.6).

The peak stress in the component is then calculated as the product of :

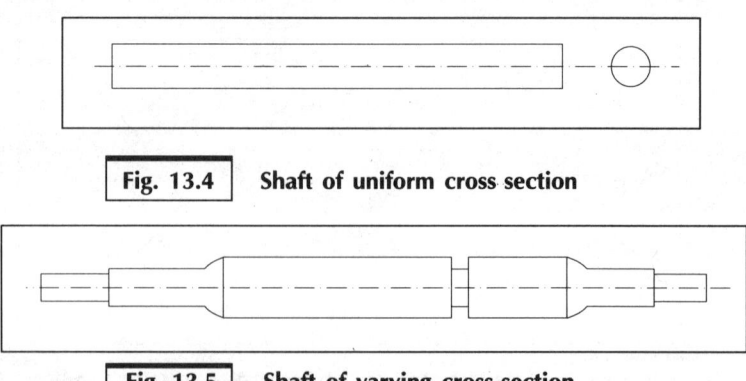

Fig. 13.4 Shaft of uniform cross section

Fig. 13.5 Shaft of varying cross section

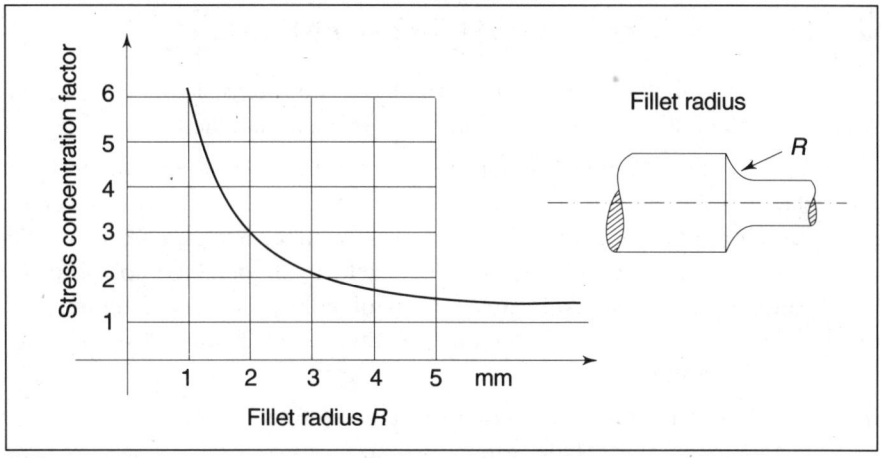

Fig. 13.6 Stress concentration factor

- stress calculated on the assumption of uniform cross-section, and
- stress concentration factor calculated on the basis of the changes in the cross section.

The detailed calculations for (a) and (b) above are beyond the scope of this book. It is only necessary to understand here that firstly, the stress concentration factor is a very important factor which comes into play in any discussion of metal fatigue, and secondly, this factor can rise from a small value like 1.5 to a very large value like 10 even due to a very small detail like the fillet radius at a change of cross section [see Figs. 13.7(a) and 13.7(b)].

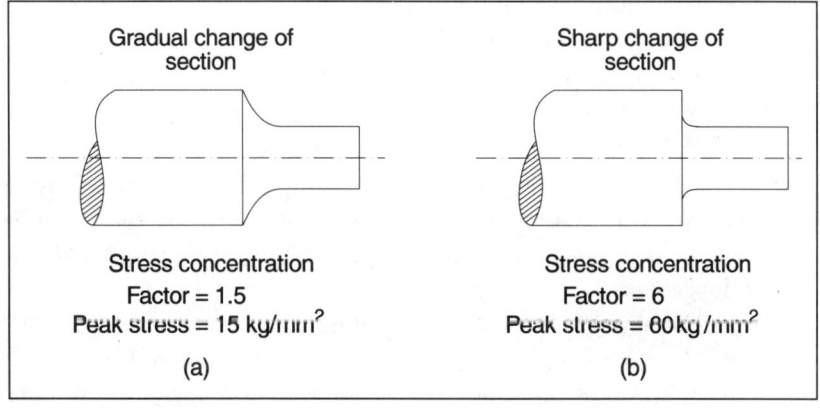

Fig. 13.7 Effect of fillet radius on peak stress

13.4 FRACTURES DUE TO METAL FATIGUE

Here, we shall discuss some of the more important factors which tend to initiate and accelerate failures due to metal fatigue.

Fractures of metal components due to phenomena other than fatigue–such as brittle fracture at low temperatures and ductile fracture under over-stress–also occur from time to time, but they are relatively rare. In the majority of cases of fracture of metal components in service, the most likely mechanism of failure is metal fatigue. It is therefore necessary to understand the basic characteristics of this phenomenon.

We can begin our discussion of metal fatigue by stating the conditions under which fatigue fractures take place. Fatigue fractures take place under the following conditions:
- The peak stress is greater than the endurance limit of the material.
- The peak stress is equal to the average stress multiplied by the stress concentration factor.
- The stress is fluctuating.
- The number of fluctuations of stress to cause failure depends on the difference between the peak stress and the endurance limit. The greater the difference, the smaller the number of fluctuations needed to cause failure.

These conditions are shown graphically in Fig. 13.8. This curve is usually known as the S–N curve. From the figure, we can draw the following conclusions:
- The endurance limit of steel is approximately 40 percent of the ultimate tensile strength.
- The number of fluctuations in stress required to cause failure when the stress is equal to the endurance limit is approximately 10 million.

The four conditions under which fatigue fractures generally take place are very important, because they also provide the key to the prevention of fatigue failures. To prevent fatigue fractures of the wire or the socket:
- The stress must be kept below the endurance limit.
- The stress concentration factor must be minimised. In any case, both stress and the stress concentration factor must not be allowed to go beyond the value provided for in the design. Although the phenomenon of metal fatigue is involved in the

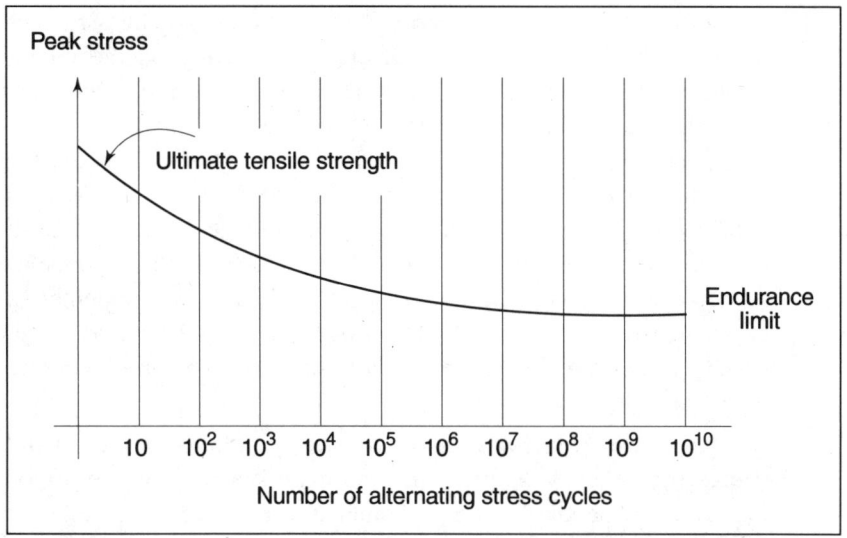

Fig. 13.8 The S-N curve showing how fatigue life (in cycles) depends on stress

failures of so many components, it is important to note that fatigue is the mechanism and not the cause of such failures.

In many cases, the fatigue crack starts as a hairline crack on the surface. This crack gradually becomes bigger and deeper, and actual failure or fracture of the component occurs only when the crack has extended by 20 to 70 percent of the cross section available.

It will be clear from the above that fatigue failures may take place at any time. If the peak tensile stress is very high and the frequency of alternation is high, the fatigue failure may take place within a few weeks or months; but if the peak tensile stress is only slightly higher than the endurance limit and if the frequency of alternation is not high, a fatigue fracture may take place only after many years of service.

The following precautions must be taken to prevent fatigue failures:
- The designer of the component must make careful calculations and ensure that the peak tensile stress at any point is less than the endurance strength of the material. While making the calculations for the peak tensile stress, he must take into account the variations in the distribution of the stresses at different points in the component—in other words, he must take into account the appropriate SCF of the component. He must also allow for

normal variations in the properties of different batches of materials and possible variations in the forces acting on the component. The lowest possible material strength and the highest possible stress must be taken into account.
- Special note must be taken of the surface finishes and the fillet radius at change of cross section, as indicated in the working drawings of the components. A poor surface finish reduces the endurance strength of a material and a smaller fillet radius at a change of cross section increases the SCF, and consequently, the peak stress. Similarly, tool marks or sharp bends increase the SCFs. All these factors are likely to lead to premature failure due to fatigue.

(A few basic principles of design are mentioned above mainly because they help to explain the rationale behind the precautions which have to be taken during manufacture and maintenance.).

With the above background about fatigue fractures, we can now see how these priniciples become relevant to our study of fractures of crimped sockets and wire strands. The precautions to be taken have already been mentioned in Chapters 8 and 9. They may be perused again after understanding the basic principles of fatigue failures as discussed in this chapter.

13.5 ALTERNATING AND FLUCTUATING STRESSES

Fatigue failures can be prevented by proper design and careful manufacture and maintenance. It is necessary for the maintenance engineer investigating fractures of metal components to be aware of the basic design aspects already mentioned, to enable him to recognise the features of failures caused by intrinsic design deficiencies. A few of these will be discussed here in more detail.

The most important feature of stresses which can lead to fatigue failures is that they must be of an alternating or fluctuating type. A steady or constant stress does not lead to fatigue failures. However, little comfort can be drawn from this fact because in most engineering equipment the stresses are alternating or fluctuating.

The distinction between alternating and fluctuating stresses will become clear from Fig. 13.9, which shows the stress values at successive instants of time.

In an alternating stress, the direction of the stress changes at every cycle. It is alternately tensile and compressive. In a fluctuating stress,

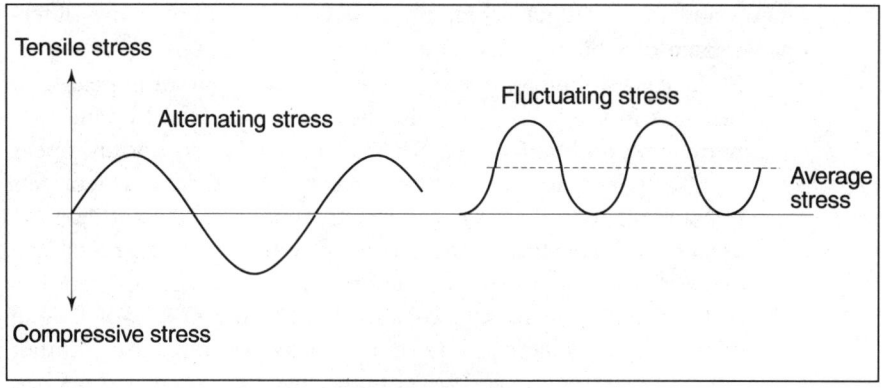

Fig. 13.9 Alternating and fluctuating stresses

the direction of the stress, remains the same, but its magnitude keeps changing. It can be seen from Fig. 13.7(b) that a fluctuating stress is equivalent to a steady stress with a superimposed alternating stress. It is the latter component or the alternating component which leads to fatigue.

As regards the magnitude of the alternating stress which can lead to fatigue fracture, the easiest way to understand the relationship is to examine the S–N curve of the metal, as shown in Fig. 13.8.

The following conclusions can be drawn from the S-N curve in Fig. 13.8.
- If the peak stress is higher than the 'ultimate tensile strength' of the material, failure occurs immediately.
- If the peak stress lies somewhere between the ultimate tensile strength and the endurance limit of the material, failure will occur after the alternating stress has completed the number of cycles indicated by the curve. The closer the stress is to the endurance limit, the larger is the number of cycles of stress before failure.

13.6 THE NATURE OF METAL FATIGUE

Just as humans get tired or fatigued after hand work, metal components also get fatigued when they are subjected to high alternating or fluctuating stresses. Again, just as humans get fatigued faster when they work more strenuously, the metal components get fatigued faster if the stress on the component is higher. The analogy between human

fatigue and metal fatigue is far from perfect. There are many differences, some of which are as follows:
- In the case of human beings, the effects of fatigue disappear after rest, but in the case of metals, the damage caused by fatigue is permanent and cumulative. The fatigue effects do not disappear by giving rest to the component. If a metal component has become fatigued due to excessive stress at any point and hairline cracks have developed, there is no way by which the metal can be restored to its original condition. The fatigued component has to be replaced. It is not possible to erase the effects of fatigue damage by annealing or any other process. (In rare cases, further damage may be arrested by grinding away the damaged surface or drilling a hole at the end of the crack, but this process is generally not recommended for ordinary engineering applications. It is used only in very large structures such as bridges as temporary relief measures, and that too only under the guidance of specialists.)
- The other important difference between humans and metals with regard to fatigue is that in case of metals, below a certain level of stress, certain metals like steels do not get fatigued at all and can work satisfactorily for an indefinitely long time if the stress is kept below a level known as the *endurance limit of the metal*. This does not happen in the case of humans, as they need periodic rest.

The simple fact to be understood clearly, therefore, is that metal fatigue can be totally prevented by ensuring that the peak stress is kept below the level known as the endurance limit of the metal. It also follows that if the stress is allowed to rise above the endurance limit, fatigue failures are bound to occur sooner or later.

Having stated the general position which applies to steels, it is also necessary to state that in the case of non-ferrous metals, there is no clear endurance limit as in the case of steels. However, by keeping the design stress sufficiently low, it is possible to ensure zero failure performance during the normal life span of the equipment.

13.7 CAUSES OF FATIGUE FAILURES

It would be relevant to ask at this stage, that if fatigue failures can be prevented by as simple a measure as keeping the stresses below the endurance limit, why do fatigue failures take place at all? Is it not possible for the designers to ensure that the endurance limit is never

Metal Fatigue

exceeded ? The answers to these questions would amount to an enumeration of the usual causes of failures through fatigue.

Ensuring that the stresses in the different components of the machine are kept within limits is actually the responsibility of the designer. Experience shows that in general, designers do calculate the stresses correctly. However, errors sometimes occur with regard to the following aspects:

- While the calculations may be correct, the results can be wrong if the assumptions regarding the forces are not correct. These should be verified, particularly if the failure rates are high indicating a design error.
- A more common error in design is with regard to stress concentration factors at changes in section or at discontinuities, apertures, etc. These aspects should be looked into if the failure rates are high and the fractures are taking place at changes in section, in the vicinity of apertures or discontinuities.
- Fatigue cracks usually start at the surface. Wherever there is a change of section or an aperture, or a discontinuity such as a slot, keyway, etc., in the component, there is a considerable increase in the local stress by a ratio which may vary over a wide range like 1.2–10 or even more. This ratio which is also known as the stress concentration factor depends on factors like radius of fillet at the change of section, etc. It is very important that the provisions regarding these aspects are followed exactly as indicated in the drawing. Any deviation in these matters can easily increase the stress by several hundred percent and lead to fatigue fractures.
- The endurance limit of a steel is reduced by factors like surface finish, and increased by surface treatment like rolling and shot peening, etc. Therefore, the stipulations in the design drawing on such matters must also be respected during manufacture, without any exception.

13.8 APPEARANCE OF FATIGUE FRACTURES

Whenever a crimped socket or wire strand fractures due to fatigue, failure usually occurs during service, when an electric current is passing through the socket or the strands. The interruption of the current leads to sparking and arcing, which causes the fractured surfaces to melt. The appearance of fatigue fractures cannot generally be studied for this reason. However, if it so happens that a fracture takes place while there is no passage of current and there is no arcing or sparking, its appearance can be seen. It will then be seen that part of the fractured surface is smooth and there is another distinctive portion which has a rough appearance. Such differences are very clearly seen in the case of fatigue fractures of purely mechanical components such as shafts.

13.9 DO'S AND DON'TS FOR PREVENTING FATIGUE FRACTURES

- Ensure that fatigue stress calculations are made during the design of all components subject to alternating stresses due to vibration or alternating thermal elongation and contraction under constraint.
- Ensure that stress concentration factors are taken into account in all stress calculations.
- Ensure that all stipulations in the drawings such as fillet radii at changes in cross section, surface finishes, surface treatments, and endurance strength of the material are fully complied with during manufacture and maintenance.
- Ensure that the vibration level on any component does not rise beyond the level assumed in the design. Ensure that the following design aspects are not allowed to change during service due to wear and tear:
  - spring characteristics,
  - clearances in bearings and suspensions,
  - points of support.
- Ensure that all fractures in service are investigated fully to determine the root causes of the fracture. Remember that fatigue is never the cause of failure; it is merely the mechanism or process of failure.

13.10 CONCLUSION

More than 80 percent of all fractures of mechanical and electrical components involve metal fatigue. When electrical conductors fracture while carrying current, the resulting arcs can cause electrical fires.

Fatigue failures occur when the peak mechanical stresses in components are in excess of the endurance limit, and when vibrations or thermal cycling cause alternating or fluctuating stresses.

A very common cause of excessive stress is the neglect of stress raisers (i.e., features which increase stress concentration factors) during design and/or manufacture. Some of these features are: sharp corners or reduced radius of fillet, tool marks, and cracks formed during metal fabrication.

Failures due to vibrations can be prevented by providing adequate support to the vulnerable components.

It is possible attain total freedom from cracks and fractures of metal components by judicious design and a few special precautions taken during manufacture, installation and maintenance.

Chapter 14

Metal Creep

In this chapter, we will discuss:

- The general definition and description of the phenomenon of metal creep.

- The conditions under which metal creep becomes significant.

- The effects of metal creep on electrical contacts which depend on mechanical force developed by threaded fasteners for low contact resistance.

- Examples of commonly used electrical equipment which are susceptible to failures caused through metal creep.

- Mechanism of electrical contact failure due to metal creep.

- Measures and precautions to be taken for preventing failures of electrical contacts due to metal creep.

14.1 INTRODUCTION

Metal creep is a phenomenon which is generally associated with elevated temperatures such as those encountered in high pressure steam boilers, pipes, and valves in modern thermal power stations. Under certain conditions, metal creep also occurs in electrical equipment operating at temperatures of about 150°C. The phenomenon of metal creep is best explained by considering a practical experiment. Imagine a weight (say 10 kg) suspended from the ceiling by means of a copper wire of length 3000 mm. Let the cross section of the copper wire be 1 mm². Thus, the tensile stress in the wire would be

Metal Creep

$10/1 = 10$ kg/mm^2. Assume that the temperature of the wire is about 20°C, which is the ambient temperature [see Fig. 14.1(a)].

Since the Young's Modulus of copper is 12000 kg/mm^2, the elastic increase in the length of the wire due to the tensile stress is calculated as follows:

Increase in length = (initial length) × (stress)/(Young's Modulus)

$$= 3000 \times 10/12000 = 2.5 \text{ mm}$$

This increase in length of the wire from 3000 mm to 3002.5 mm under the effect of the tensile stress is constant with respect to time. Even if the wire is held under the same tensile stress for many years, there would be no further increase in the length; and if the stress is removed after all that period, the wire will regain its original length of 3000 mm. This assumption is actually quite true, but only if the temperature of the wire is less than 135°C.

If in the above example, the length of the copper wire was 3000 mm and its temperature was 200°C to start with (say by passing an electric current through it), the increase in length on applying the tension of 10 kg/mm^2 initially, would be the same as before, i.e., 2.5 mm, but this time it would not remain constant with time. The wire would continue to increase in length at the rate of about 3 mm per 1000 hours. If the tension is reduced to zero after 1000 hours, the wire will not regain the original length of 3000 mm. The increase of 3 mm will be a permanent increase, while the increase of 2.5 mm, which took place immediately on applying the tension will be reduced to zero. The new length of the wire will be 3003 mm [see Fig. 14.1(b)].

Note. It may be noted that we have not considered here another phenomenon, viz. thermal expansion. This is quite independent of the above phenomena. The second experiment started with a wire of length 3000 mm at 200°C.

The permanent increase in the length of the copper wire under the combined effect of tensile stress, high temperature and the passage of time is due to the phenomenon of metal creep. In this example, we have considered a thin long wire in tension; but the same principle applies even to a small block in compression. The magnitudes will be smaller, but the percentage strains would be the same. A block of thickness 3 mm will, under a compressive stress of 10 kg/mm^2 reduce in thickness by 0.0025 mm. At 200°C, the creep strain would be 0.003 mm.

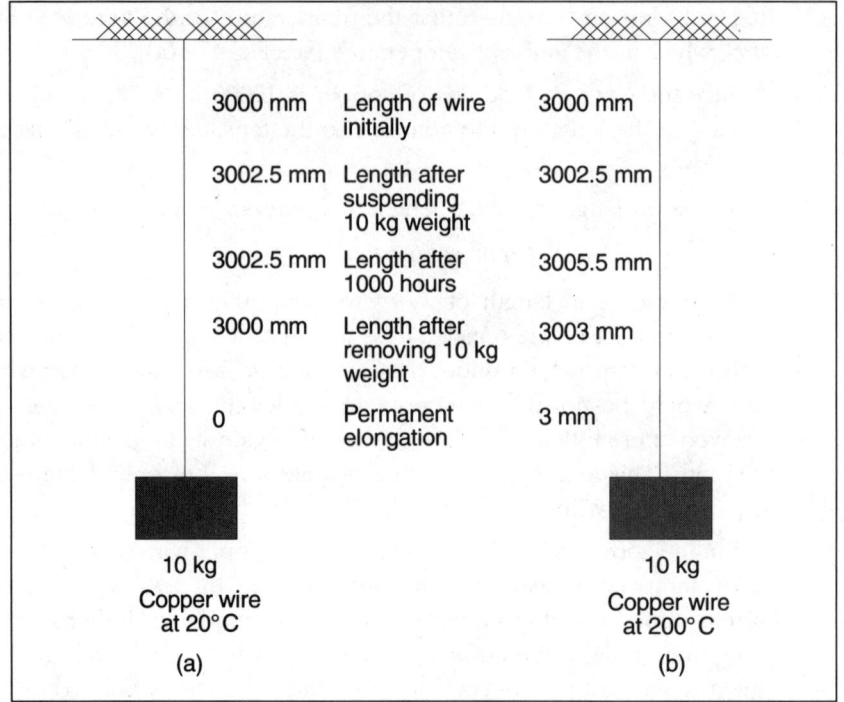

Fig. 14.1 Effect of time on elongation of copper wire at 20°C and 200°C

Figure 14.2 shows a set of curves which depict the creep behaviour of metals.

The following features of metal creep may be seen from Fig. 14.2:
- With the passage of time, the creep rate decreases initially. Later, it becomes constant up to a point and then starts rising rapidly again. These three zones are known as primary, secondary and tertiary creep zones. Entry into the tertiary zone signals approaching failure.
- The creep rate is higher at higher stress levels.
- If tests are carried out at different temperatures it is observed that the creep rate is higher at higher temperatures.

14.2 EFFECTS OF METAL CREEP ON ELECTRICAL CONTACTS

It may be noticed that the increase in length due to metal creep is not only very small, but also very slow. However, even this small and slow

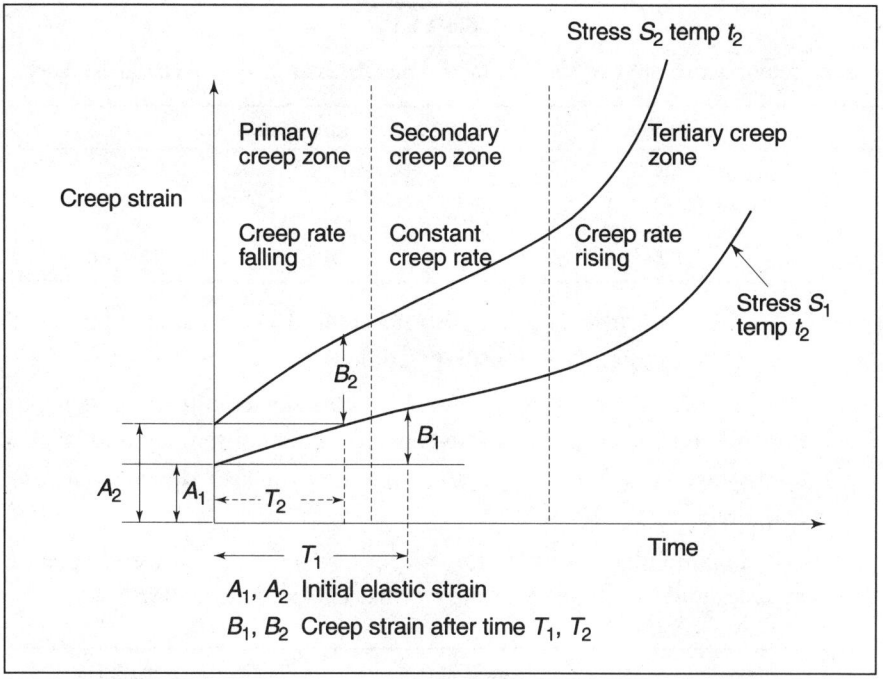

Fig. 14.2 Effect of time on elongation of copper wire at 200°C

deformation is the root cause of some electrical failures—failures which have the potential of growing into fires. Metal creep is a phenomenon which sometimes affects the reliability of electrical connections involving pressure contacts. Failures due to metal creep are not very common, but they do occur from time to time, and it is desirable to understand the mechanism of such failures and to recognise the conditions under which failures of this type can occur.

Metal creep can be described as the continuous deformation of metals which takes place under certain conditions as follows:
- Metal creep occurs when the temperature of the metal is higher than the creep temperature limit of the metal.

This creep temperature limit (CTL) depends on the melting point (MP) of the metal. (CTL = MP/3 −182°C approximately). Data regarding some of the common metals used in the electrical industry are given in Table 14.1.

| Table 14.1 |

Creep temperature limit of some common metals used in the electrical industry

Metal	Creep Temperature Limit (°C)
Steel	270
Copper	135
Aluminium	7
Lead	−70

- For creep to occur, it is also necessary for the metal component to be under either tensile or compressive stress.

The time rate at which a metal component continues to deform, depends on the difference between the operating temperature of the component and its creep temperature limit. It also depends on the magnitude of the stress.

One particular creep rate for copper is given here to give an idea of its magnitude in relation to the normal elastic deformation:

Stress (kg/mm^2)	Creep Rate at 200°C (percent/1000 hr)	Elastic Deformation (percent)
10	0.1	0.08

It would be seen from the above comparison that creep rates can be of the same order of magnitude as elastic deformation. If the stress levels are higher or if the temperatures are higher, the creep rates are even more predominant. The relationship between temperature or stress and the creep rate is not linear. As temperature or stress rises, the creep rate rises more and more rapidly. Further, creep rates vary from metal to metal.

Metal creep, though apparently very slow and very small in magnitude, is a real source of problems when internal forces produced by elastic deformation of the components in assemblies are vital for the performance of the equipment. It may be recalled that the quality of electrical contacts is extremely sensitive to the internal assembly forces between the contacting surfaces. Relaxation of these forces as a result of metal creep is the root cause of some electrical contact failures. Therefore, metal creep is important in the context of electrical connections.

14.3 FAILURES OF ELECTRICAL CONNECTIONS DUE TO CREEP

As far as bolted connections of copper components are concerned, the operating temperatures are usually well below the creep critical temperature of copper and creep is not a problem with copper components. However, as may be noticed from the Table 14.1, lead and aluminium have creep critical temperatures which are lower than the usual operating temperatures. Metal creep can cause problems when these metals are used. Some examples are given below:

- Bolted connections of aluminium busbars or sockets tend to relax or loose their contact force due to the gradual squeezing of the busbars under compression. This increases the temperature rise, causes a further increase in the creep rate and leads eventually to failure of the joint.
- Parallel clamps and bolted connections on outdoor aluminium conductors are more vulnerable to failures caused by creep because their operating temperatures are usually higher due to exposure to sunlight.
- In household wiring with aluminium wires, bad contact, sparking, arcing and overheating at terminals is usually due to metal creep.
- The terminals of lead-acid batteries are usually made of lead. Bolted connections on lead terminals have a tendency to develop loss of contact force and consequent increase in the contact resistance.
- Soldered joints which use soft solders (lead/tin) are prone to fail in service basically due to metal creep if the solder is allowed to bear any mechanical stress.

The mechanism of failure due to creep can be understood by considering the case of a simple bolted joint of two aluminium busbars. Such a joint is shown in Fig. 14.3.

It would be seen that a part of the aluminium busbar between the head of the bolt and the nut would be under compression, while the shank of the bolt would be under tension. There would be some elastic compression in the busbar and some elastic elongation in the shank of the bolt. Assuming that the bolts and nuts are made of steel, there would be no creep in the bolt/nut because the operating temperature would be much below the creep temperature limit of steel.

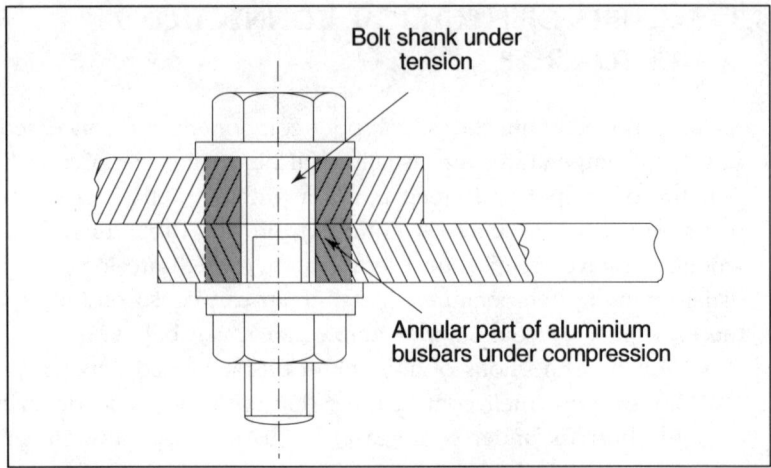

Fig. 14.3 Bolted joint of two aluminium busbars showing zones under compression and subject to metal creep

However, the aluminium bars would be operating above their creep temperature limit and this would cause the aluminium under compression to reduce continuously in thickness as a result of metal creep. This would be a gradual and continuous, but very slow process. Some of the elastic deformation of the aluminium busbar would be replaced by the permanent creep deformation. There would thus be a relaxation or reduction in the residual elastic deformation and a consequent reduction in the contact force. The relaxation in the contact force due to creep strain replacing part of the elastic strain is shown in Fig. 14.4.

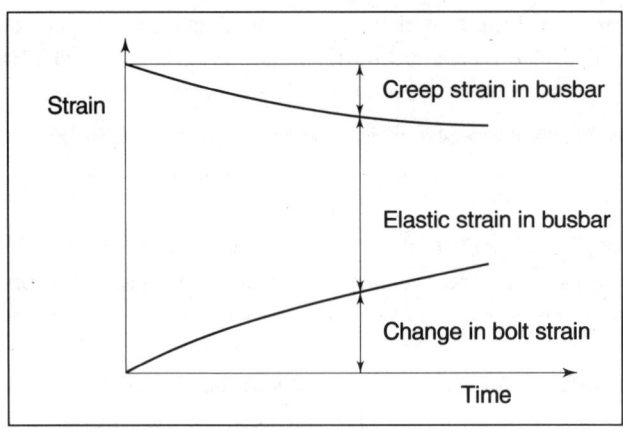

Fig. 14.4 Effect of creep in a bolted joint

The reduction in contact force would, in turn, lead to an increased contact resistance, an increase in operating temperature and an increased creep rate. A vicious cycle similar to that shown in Fig. 6.6 would be established, which would now end only in the arcing and burning of the joint, unless timely corrective action is taken by re-tightening the bolts and nuts. In this vicious cycle, creep would be an added aggravating element (see Fig. 14.5).

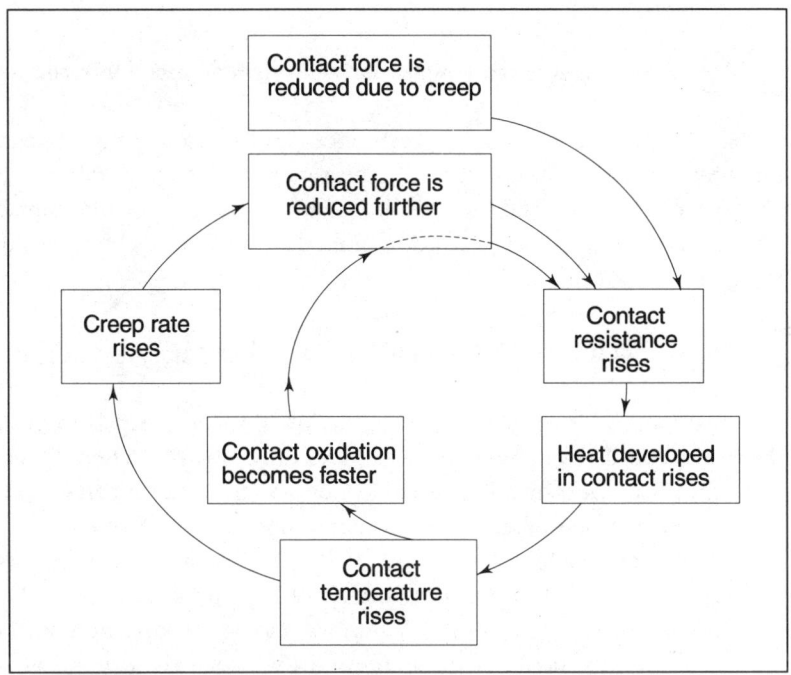

Fig. 14.5 Vicious cycle in mechanism of failure of bolted joint between aluminium busbars due to creep

The mechanism of failure in a simple stud connection (Fig. 14.6) such as is used in a wall socket or switch in a domestic installation is very similar, although the dimensions are much smaller. In Fig. 14.6, the aluminium wire A is clamped by the stud s, which provides the contact force necessary for a good electrical contact. When the stud is tightened, there is a compressive stress in the stud, as also in the aluminium wire. The terminal T develops tensile stresses which balance the compressive stresses in the stud and the wire. As the stud and the terminal are made of brass, they are not subject to creep, but the aluminium wire will gradually relax its compressive stress due to a growing creep deformation of the wire. The reduction in the contact

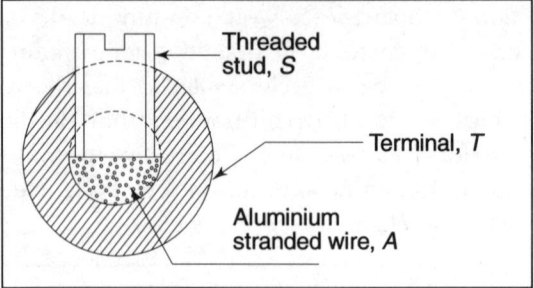

Fig. 14.6 Failure of screwed terminal for aluminium wire due to creep

force will increase the contact resistance, and consequently increase the heat developed in the contact. The same vicious circle referred to above will develop. The end result would be burning of the contact and the wire, unless the process is nipped in the bud by retightening the stud periodically.

14.4 PREVENTION OF FAILURES DUE TO METAL CREEP

The best method for preventing failures due to metal creep is to avoid the use of metals like aluminium, lead and tin. However, when the use of such metals becomes unavoidable due to any reason, certain precautions have to be taken. These are as follows:

- If any bolted joints are used, Belleville washers (disc springs) should be used in the force circuit of the connecting bolts. The design of the disc spring assembly should be matched to the force requirements of the joint, and their total deflection, when fully tightened, should be at least 10 percent of the total thickness of the aluminium under compression (see Fig. 14.7).
- All bolted or screwed joints in lead or aluminium conductors or busbars should be tightened and overhauled periodically.
- If tin-lead solders are used for any joint, it should be ensured that the conductors are secured in place mechanically, by twisting or bending the wire around the lug, so that there is no mechanical stress on any part of the solder. It may be recalled that tin-lead solder has a life of only 1000 hours at a stress as low as 0.4 kg/mm^2 and temperature of 100°C.

There are other possible causes of failures of aluminium joints. These are oxidation, bimetallic action and inadequate tightening or bolts/nuts or screws. In many cases, metal creep may be only a

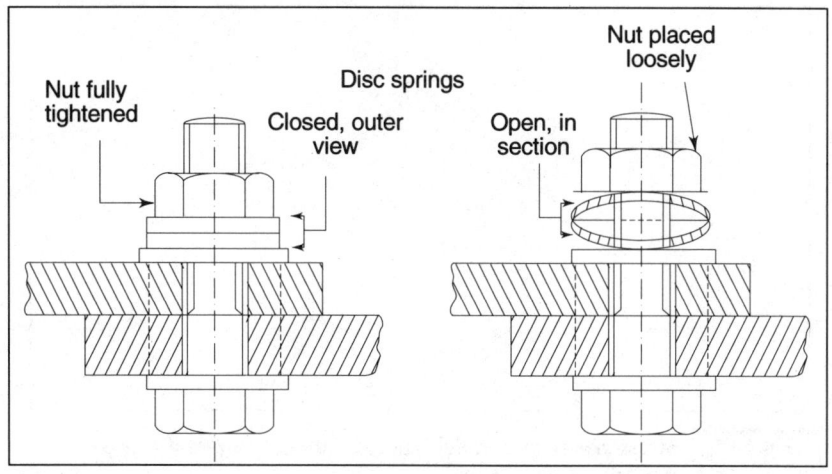

Fig. 14.7 Use of disc-springs to prevent bolted joints, between aluminium bars, due to creep

contributory factor. It is therefore necessary to investigate all these possibilities also, when faced with failures of aluminium conductor joints.

It has been mentioned that metal creep does not create any problems for connections made with copper wires, sockets and busbars. This is true only for ordinary connections at normal ambient temperatures. Metal creep can become a source of problems even with copper conductors inside ovens where the temperatures are above 150°C, or in electrical machines like traction motors where the copper temperatures are likely to be higher than 140°C. Bolted, crimped or soldered electrical connections are unsuitable under such conditions. Only brazed or welded joints are suitable.

14.5 FAILURES OF PARALLEL CLAMPS

Failures of parallel clamps used on overhead lines, could be due to inadequate tightening of the fasteners initially during installation.

They could also be due to metal creep in the aluminium wires and clamps (Fig. 14.8).

In the parallel clamp shown in Fig. 14.8, the outer faces of the clamp are under compression caused by bending, and the inner faces are under tension caused by bending. The effect of metal creep is to cause relaxation of these stresses, which is equivalent to a reduction in the bending of the clamp plates. This leads to a reduction in the

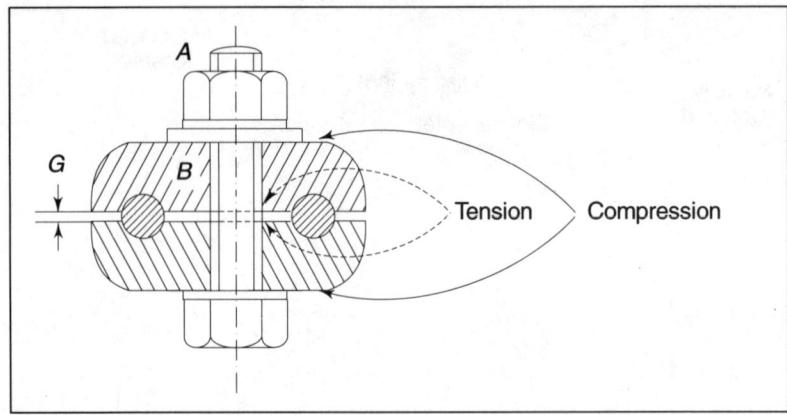

Fig.14.8 Stress-zones in parallel clamps. Subject to metal creep if made of aluminium

contact forces and the usual process of heating and eventual failure results.

There is one more trivial but common cause of failure of parallel clamps, viz. contact between the two halves of the clamp due to dimensional errors in the grooves. If this happens, the contact force between the conductors and the clamps would be reduced and the usual failure mechanism would be set into action. The gap G shown in Fig. 14.8 is vital for reliability.

Similar connections are sometimes used in electrical equipment too–for example, traction motor terminals. These are usually made of tinned copper and creep is not a problem, but the gap G mentioned above is very important.

14.6 DO'S AND DON'TS FOR PREVENTING FAILURES DUE TO METAL CREEP

- Do not use, as far as possible, aluminium or its alloys in applications where pressure contacts are provided.
- If the use of aluminium in pressure contacts is unavoidable, re-tighten the pressure contacts periodically. The intervals may be determined by trial and error.
- If aluminium pressure contacts are unavoidable, use disc-springs or Belleville washers under the nuts which provide the contact force. This will avoid the need for frequent retightening of the nuts.

- Avoid the use of soldered joints even when the conductors are made of copper. Use crimped or bolted joints instead. If the operating temperatures are likely to exceed 135°C, use brazed or welded joints. If soldered joints are unavoidable, ensure that the wires are mechanically secured by twisting or bending around the lugs before soldering.

14.7 CONCLUSION

Metal creep is a phenomenon which occurs in all metals under mechanical stress when their operating temperatures are higher than their respective creep temperature limits.

The effect of metal creep is a continuous increase in the deformation or strain caused by a mechanical stress.

Creep rates are apparently small and slow, but their magnitudes are comparable to elastic deformations and their cumulative effects over years of service are sufficient to cause failures of electrical contacts which depend on internal elastic forces for good electrical contact.

Failures of electrical contacts through the effects of metal creep can be prevented by:
- Avoiding the use of aluminium conductors.
- Use of Belleville washers under nuts.
- Periodical retightening of fasteners.

Chapter 15

Electrical Contact Resistance

In this chapter, we will discuss:

- Elementary theory of electrical contacts sustained by mechanical force between the conductor surfaces.

- Three common types of such contacts, viz. bolted connections, plug/socket connectors and crimped sockets.

- Factors which influence the electrical contact resistance of such contacts.

- The mathematical relationship between mechanical force and electrical contact resistance and the theoretical basis for it.

- Explanation for mechanisms of failures of electrical contacts in terms of the force/resistance relationship.

- The procedures and precautions during the design, manufacture and installation of electrical connections which depend on mechanical force.

- Additional precautions, viz. elimination of dust for ensuring reliability of light, i.e. low-force contacts.

15.1 INTRODUCTION

There are many points in electrical connectors where the current has to pass from one component to another through a surface contact between the two components. Examples of such points are:
- between the strands of a cable and the barrel of a crimped socket,

- between the palm of the socket and the terminal of the electrical equipment,
- between the plugs and the sockets of a multicore coupler,
- between two busbars connected together by bolt/nut fasteners,
- between two components in an electrical device such as a circuit breaker, e.g., between the contact tip and the contact carrier, or between the moving contact and the braided flexible connector.

There are many other points inside electrical equipment where the current passes from one component to another through surface contact. The general principles which apply to all these electrical surface contacts are the same, but this chapter is specifically about the electrical contact resistance of such surface contacts in electrical connectors.

Contrary to what one might expect, the electrical contact resistance between two conductors in contact with each other does not depend on the nominal area of contact. It depends only on the mechanical force between the two conductors, apart from certain mechanical and electrical properties of the material of the two conductors.

15.2 IMPORTANCE OF ELECTRICAL CONTACT RESISTANCE

Whenever two electrical conductors are held together with a view to enable the passage of electric current from one component to the other, there is some additional electrical resistance at the point of surface contact. The effect of this contact resistance is to cause a voltage drop at this point and also, more importantly, to cause generation of heat at this point of contact. This heat increases the operating temperature of the components.

If there is any deficiency in the design, manufacture or maintenance of the electrical contacts inside an equipment, the temperature can rise to levels at which insulating materials may be damaged. Electrical fires can also originate due to such deficiencies. It is very important, therefore, to ensure that the operating temperatures are kept within limits and for this purpose, it is necessary to understand clearly the factors which influence the electrical resistance of force contacts.

15.3 FACTORS WHICH INFLUENCE ELECTRICAL CONTACT RESISTANCE

The electrical resistance between two components in contact with each other depends on the following factors:

- The normal (i.e., perpendicular to the surface) force between the two components, i.e., the force with which the two components are held together. (This force is usually provided by threaded fasteners or by springs.)
- The resistivities of the metals at the point of contact.
- The yield point of the metals at the point of contact.
- The cleanliness of the contacting surfaces.

The most important of these four factors is the first, i.e., the normal force between the components. There is a definite relationship between the force and the contact resistance. It may be noted that the relationship is not linear. There is considerable variation between observations of contact resistance even when the force and all other variables are kept constant. Even so, it is very clear that the higher the contact force, the lower is the contact resistance. See Fig. 15.1 which shows the contact resistance between two copper flats, as a function of the force between the flats.

That the contact resistance depends on the resistivity of the contact material is to be expected; but the reasons for its dependence on the yield point of the contact material may not be very clear at first. When two metal flats of area A are pressed together as shown in Fig. 15.2(a), it appears as if the area of contact is A; but this is only apparently so. The real contact area is much less because of surface irregularities.

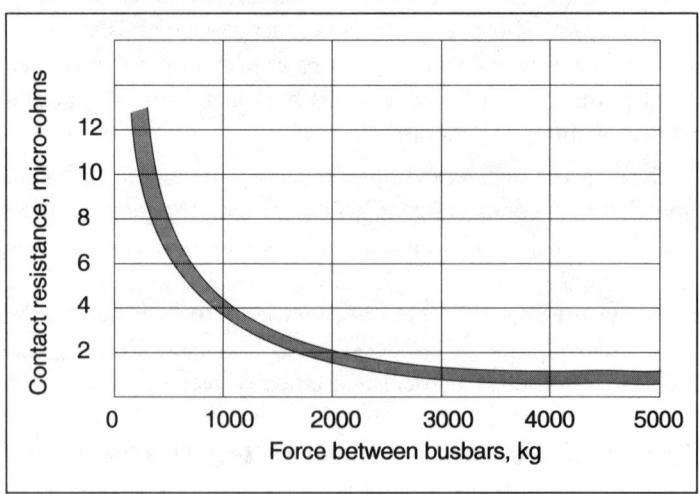

Fig. 15.1 Contact resistance between two copper flats as a function of the contact force between them

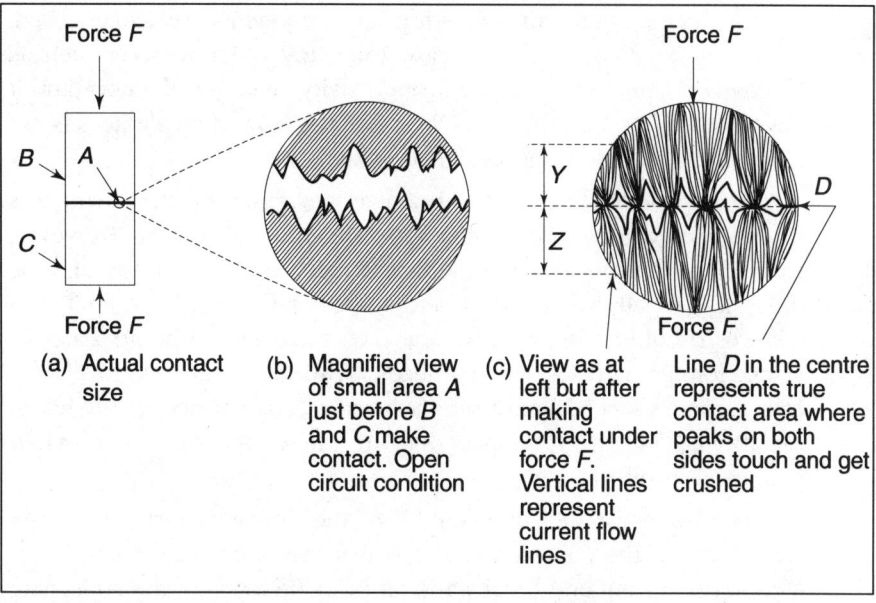

Fig. 15.2 (a) apparent contact area (b) magnified view of ridges and valleys on contacting surfaces before making contact (c) magnified view showing true contact areas formed after applying force

A highly magnified sectional view of the contact would be as shown in Fig. 15.2(b). Actually, the surfaces are smooth by our normal standards and the ridges and valleys (or asperities) seen in the magnified view have an amplitude of less than two microns. Physical contact takes place at only a few points. The rest of the space between the surfaces will be filled with air which is a good insulator. If a force is now applied between the two flats, the ridges in contact will get crushed and the area of contact will increase until this area multiplied by the yield point of the metal equals the force applied. Electric current will spread out from these small areas of contact into the whole conductor as shown in Fig. 15.2(c). Thus the true area of contact is equal to force/(yield point). As the contact resistance will obviously depend on the true area of contact, it follows that it will depend not only on the force, but also on the yield point of the material.

More detailed explanations regarding this aspect are beyond the scope of this book. Those interested may refer to books on electrical contacts. In any case these factors are of importance only to the designer of the equipment. Those concerned with the manufacture and maintenance of electrical equipment have only to see that the metals specified in the drawings for the components are actually used. This they have to do in any case for many other reasons such as mechanical strength, electrical conductivity, etc. More important it has to be ensured that the force between the contacting surfaces is not allowed to fall below the designed value.

The dependence of electrical contact resistance on the cleanliness of the contacting surfaces also seems to be self-evident. However, cleanliness of the contacting surfaces is not as important as may be thought, specially when the contact force is above a certain limit. The surfaces should be reasonably clean, i.e., there should be no scales or visible foreign particles. It is not necessary for the surfaces to be chemically clean. In fact, a film of oil on the surface is not necessarily harmful. Bolted electrical connections are quite effective even when immersed in oil.

The reason for this insensitivity of the contact resistance to the cleanliness of the contact surfaces is that the force between the contacts squeezes out and breaks any oil or oxide films on the contacting surfaces, thereby establishing metal to metal contact between the surfaces of the two components.

When the contact force is low, as in low-current relay or auxiliary contacts, it is desirable and often essential to enclose the contacts in dust-tight covers. Where this is not practicable, the entire room may be sealed and pressurised with clean filtered dust-free air. The precautions are necessary to avoid bad contacts due to dust. When the contact force is low, dust or filmy deposits may not be dislodged when the contacts are closed.

In some contacts, a wipe-action is provided by suitable mechanical design of the contact and its carrier. In this arrangement, after the contacts touch, there is a small linear movement or wiping action which helps to dislodge any foreign material and ensure direct contact between the metals of the moving and fixed contacts. Contacts of this type are more reliable.

Some readers may have noticed and wondered about the absence of 'contact surface area' from amongst the factors which influence the electrical resistance of contacts. This is not an inadvertent omission. It is a fact that electrical contact resistance does not depend on the surface area of the contact. A brief explanation for this peculiar result is given in Section 15.4. The true area of contact where direct metal to metal contact is established is usually less than 5 percent of the nominal contact area, and it depends only on the contact force and the yield point of the material.

In short, therefore, the most important point to be ensured for preventing failures of electrical contacts due to over-heating, burning, and arcing is the force between the contacting surfaces. If this force is provided by threaded fasteners, they must not only be tightened fully to begin with, but also retained in that state by the use of appropriate locking devices. Where the contact force is developed by springs, the following points should be ensured:

- The dimensions of the spring must comply with the drawing/design.
- The deflection of the spring in the final assembly must be within the limit specified in the drawing/design.
- In the case of plug and socket connectors where the springs are not accessible, the force required to pull out the pin from the socket should be measured. It should be ensured that this pull out force is within the specified limits.

There are other factors too which influence the reliability of the contacts. In particular, contacts which open on load have to be specially designed to interrupt the current and to deal effectively with the arcing which takes place on opening the current; but these are well known and visible aspects. The speed of opening, arc blow out arrangements, contact gap when open, and the quality of the material of the contacts are all important details which must be maintained as in the original proven design. Any changes can create problems.

15.4 CONTACT THEORY IN BRIEF

There is a great deal of literature on the subject of electrical contact resistance. Here, we shall only explain why or how electrical contact resistance is independent of the nominal contact area and so dependent on the contact force.

Consider two copper rods b and c touching each other as shown in Fig. 15.3. In this example:

The resistances of bars b and c are given by the standard formulae for resistance as shown below:

$$R_b = R_y \times L_b/A_b \text{ ohm}$$
$$R_c = R_y \times L_c/A_c \text{ ohm}$$

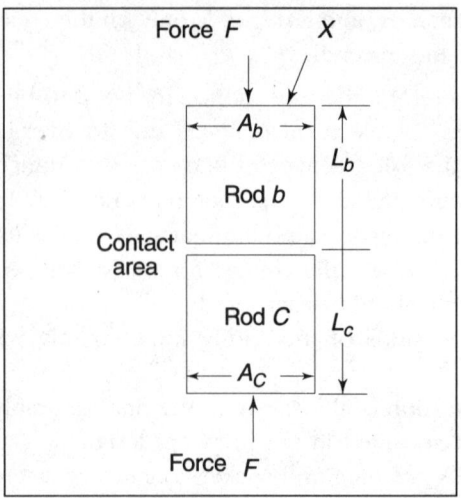

Fig. 15.3 Contact resistance as distinct from resistance through metal

where

R_b and R_c are the resistances of the bars b and c

R_y is the resistivity of the material

L_b and L_c are the lengths of the bars b and c

A_b and A_c are cross-section areas of bars b and c

R, the contact resistance between the rods b and c, depends on the contact force F.

Total resistance = $R + R_b + R_c$

The resistances of the two bars are thus inversely proportional to their cross-sectional areas. However, the contact resistance R *between* the two bars at the point of their contact is independent of the cross-sectional area of either bar. It depends only on the compressive force between the two bars and the materials of the two bars. Variation in the cross-sectional area of the two bars will have no effect on the contact resistance if the force and the material are kept constant.

Even when the contact surfaces are smooth and polished, if we look at them through a powerful microscope, we will see ridges and valleys as shown in Fig. 15.2(b). Hence, when they touch each other, as shown in Fig. 15.2(c), they touch at only a few points. The rest of the space between the surfaces is filled with air which is a bad conductor of electricity. The ridges get flattened until the total area of contact is equal to the force between the contacts divided by the yield point of the material. Electric current spreads out from these small contact areas into the rest of the conductor. The increase in resistance of the bars in these zones Y and Z, due to the restricted flow of current, appears as the contact resistance. This explains why the so-called contact resistance depends on the force between the bars and not on the nominal cross-sectional areas of the bars.

The theory of contacts could be studied further, but for all practical engineers, it is only necessary to remember that the contact resistance between any two conductors depends only on the force between them. If the force is reduced, the contact resistance rises and so does the heat developed in the contact. The contact starts overheating and after 90°C, oxidation of copper takes over the failure process.

15.5 DO'S AND DON'TS FOR PREVENTING CONTACT FAILURES

- In the case of electrical contacts which depend on threaded fasteners for the contact force, ensure that the fasteners are tightened fully to start with, and that they are maintained in the tight conditions by the use of locking devices and Belleville washers as may be needed.
- In the case of electrical contacts which depend on springs for the contact force, ensure that the spring characteristics and the spring deflection do not get reduced during service.

Both the above types of electrical contacts have one thing in common. The quality of the contact, i.e., the electrical resistance of the contact, depends mainly on the contact force between the contact surfaces.

- Ensure that the contact surfaces of light contacts are kept free from dust by suitable covers. Examples of light contacts are found in small relays where the currents, and consequently, the contact forces are usually very small.
- Ensure that the contact materials are the same as stipulated in the design and proven by tests. Ensure also that the contact gap when open, arc blow out arrangements, and opening speed are all maintained as in the original design.
- Verify that the contact temperatures when carrying full load current do not ever rise above 85°C.

15.6 CONCLUSION

Bolted connections, plug/socket connectors and crimped sockets have one feature in common: the electrical contact resistance between the contacting surfaces is determined by the mechanical force between them. If this force is maintained above certain limits, the contacts remain reliable. If not, the contact resistance and the contact temperature keep rising until, eventually, the contacts fail through burning and arcing.

All procedures and precautions during design, manufacture and installation are directed towards the very important requirement of ensuring the required minimum mechanical force between the contacting surfaces.

Light contacts such as those in relays have an additional requirement, viz. elimination of dust between the contact surfaces.

Zero failure performance can be expected if the simple precautions mentioned above are taken.

Chapter 16

Shrinkage of Non-Metallic Materials

In this chapter, we will discuss:

🔌 The two failure modes—failures of terminals due to shrinkage of insulated laminated boards, and failures of transformer coils due to shrinkage of the paper-based insulating layers and spacers.

🔌 Some other types of equipment failures, caused by the shrinkage of non-metallic materials, such as:

- Slackening of insulated bolts.
- Failure of reactor coils.
- Failure of traction motor field coils.
- The measures to be taken to prevent equipment failures due to the shrinkage of non-metallic insulating materials.

🔌 Design features of reliable terminals, which remain immune to the effects of shrinkage of insulating materials.

🔌 16.1 INTRODUCTION

The majority of laminates, plastic boards and other non-metallic insulating materials have a common property, viz. shrinkage under the effect of pressure and/or heat. The shrinkage is not very high in absolute terms. It may be in the range 0.5 to 5 percent of the thickness of the sheet. Small as this shrinkage may appear to be, it is high enough to lead to serious failures of electrical equipment if the designer does not take it into account and make suitable allowance for it.

There are two different mechanisms of failures arising out of shrinkage of insulating materials. In the first, shrinkage allows relative movement between components which leads to mechanical damage of the insulation. In the second, shrinkage reduces the contact force, increases contact resistance, causes overheating and thermal damage of the insulation.

This property of shrinkage of non-metals is very important in the design of transformers, reactors and similar electrical equipment in which large copper windings are stacked together. Shrinkage of non-metals is also important in the design of terminals, and it is this application which is discussed first. It is proposed to discuss a common defect in the design of terminals where the shrinkage of non-metals is ignored by the designers. It is of very great importance because the consequences of such design defects are often serious fires involving heavy losses.

16.2 DEFECTIVE DESIGN OF TERMINALS

The most common type of design defect relevant to the property of shrinkage of insulating sheets is in the design of bolted terminals. Figure 16.1 shows one such defective design.

It can be seen that in this terminal design, there is a common threaded fastener which provides the contact forces for the contact between the sockets and the fastener, and also the force to hold the entire assembly on an insulating board.

There are some other design defects in Fig. 16.1.
- the current passes through the bolt shank and also the nut,

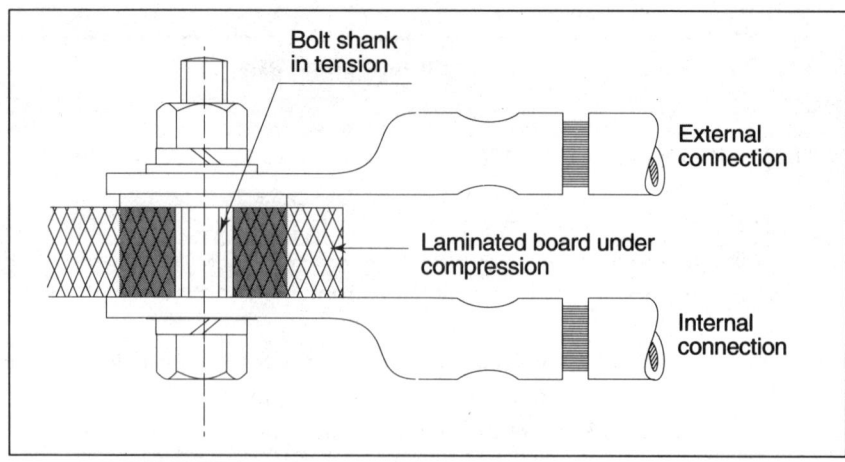

Fig. 16.1 Bolted terminal showing a defective design

- there are five pressure contacts—internal socket to bolt head, bolt to nut, nut to spring washer, spring washer to flat washer flat washer to external socket.

The first defect is not too serious if the bolt is made sufficiently large in cross-section. The second defect would be acceptable if all five pressure contacts could be made totally reliable. The real problem arises out of a more serious and insidious seed-defect in the whole arrangement. It can lead to a failure of the terminal after some time even if on the day of installation, the terminal functions quite satisfactorily. The root cause of failures in this arrangement is the shrinkage of the insulating board which takes place under the effect of the pressure exerted by the fastener.

The mechanism of failure is as follows: When the nuts/bolts are fully tightened at the time of installation, the bolt shank develops considerable tension. This tension provides the contact force which is so necessary for ensuring that the electrical contact resistances between the five pairs of pressure contacts are low. The temperature of the terminal arrangement is then within limits (less than 90°C).

However, the contact force is greatly reduced when the insulating material shrinks under the influence of pressure. The elastic strain in the bolt shank is reduced by the same amount as the shrinkage in the board. There are corresponding reductions in the bolt stress, the tension in the bolt and the contact force. This results in an increase in the electrical resistance of all the five contacts between the components listed above. The temperature of the fastener increases beyond the permissible limit, thereby initiating other degradation phenomena such as oxidation and perhaps even metal creep. A vicious cycle is thus established, which usually culminates in the charring and ignition of the combustible laminate. This may result in a fire which could spread far beyond the equipment itself. Consideration of the actual magnitudes of stresses and deformations may clarify the issues further.

When the terminal bolt which is usually made of mild steel is tightened fully, the tensile stress in the shank is of the order of 18 kg per mm^2, which is somewhat less than the yield point of mild steel. Since the Young's Modulus of steel is 18,000 kg per mm^2, the elongation of the bolt shank when fully tightened is about 0.1 percent of the length of the shank. It is, therefore, clear that a shrinkage of the order of only 0.1 percent of the thickness of the insulating laminate (in Fig. 16.1) would cause the tension in the bolt, and consequently the contact forces between the sockets and the bolt/nut, to disappear completely.

Overheating and burning of the terminal is likely to occur long before the contact forces are reduced to zero, because substantial contact forces are essential for keeping the contact resistance and the voltage drop within safe limits.

Design defects of this type in terminals are not as rare as might be assumed; and where such defects exist, failures and fires are certain to occur sooner or later. Most of the electrical fires which are commonly ascribed to 'electrical short circuits' are actually due to failures of electrical terminals or connectors.

The problems mentioned above are more likely to occur in the case of high current terminals. Overheating, arcing and burning may not occur in light, control circuit terminals, but even there, intermittent open circuits can occur if the contact forces are reduced by shrinkage of insulating materials in the force circuits of terminal bolts.

16.3 CORRECT DESIGN OF TERMINALS

Figure 16.2 shows a design in which the defects mentioned above have been eliminated. In this design, the force circuit of the forces between the sockets does not contain any non-metallic material. Further, there is no current in the fastener shank as there is direct contact between the internal and external sockets. Thus there is only one pressure contact in this system instead of five, as in Fig. 16.1. There is a separate force circuit for the force which holds the terminal in position on the terminal board.

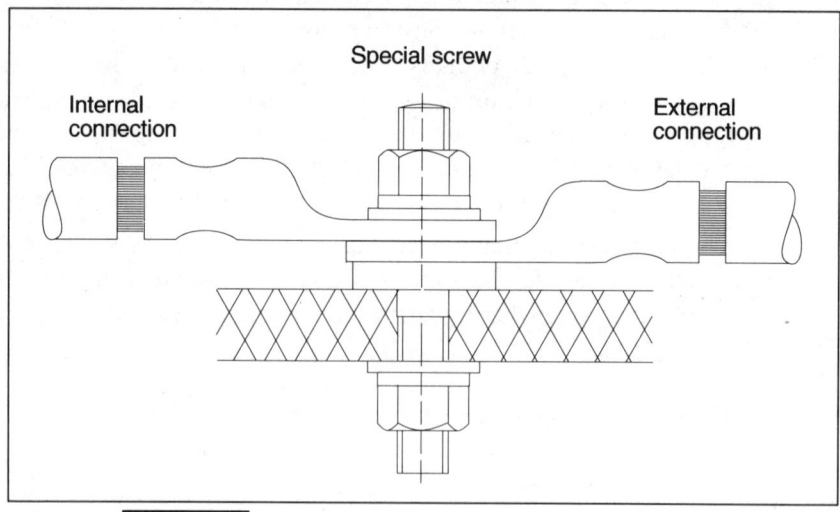

Fig. 16.2 Corrected design of bolted terminal

Although the insulating board does shrink even in this arrangement, this shrinkage has no effect on the electrical contact resistance between the sockets. The reduction of force which holds the terminal on the insulating board is not harmful in any way. Such a terminal design is very reliable. A zero failure rate can be expected, particularly if it is correctly dimensioned and adequately tightened during installation, so as to operate at a temperature well below 90°C.

There are several other reliable designs of terminals and all these have one feature in common—that the force circuit of the contact force is entirely metallic and there are no non-metals in that force circuit. Figure 16.3 shows a design where the terminal block is moulded into the board. In this case too, there is direct contact between the internal and the external sockets and the force circuit for the contact force is entirely metallic.

Yet another reliable design is shown in Fig. 16.4. In this design, ceramic insulators are used to support the terminal blocks which are moulded into the insulators. Contact between the internal and the external cable sockets is direct. The force circuit is again entirely metallic. Terminals of this type are used in many rotating electrical machines operating at voltages up to 1500 V.

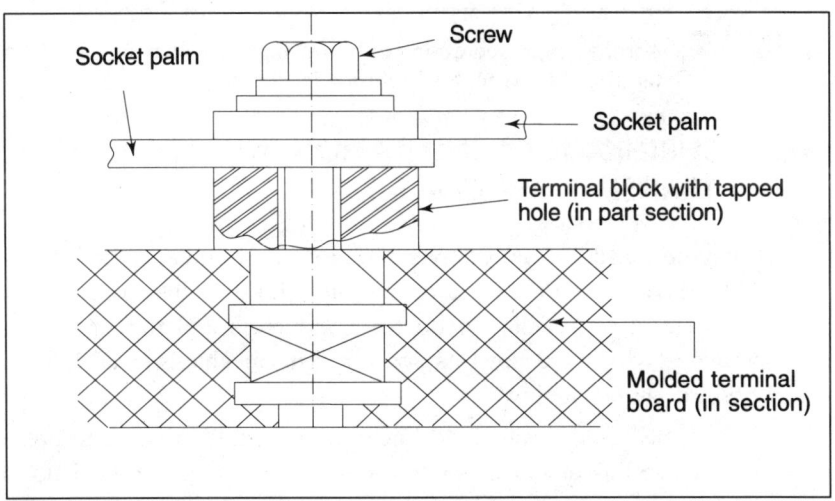

Fig. 16.3 Design of bolted terminal showing terminal block molded into the board

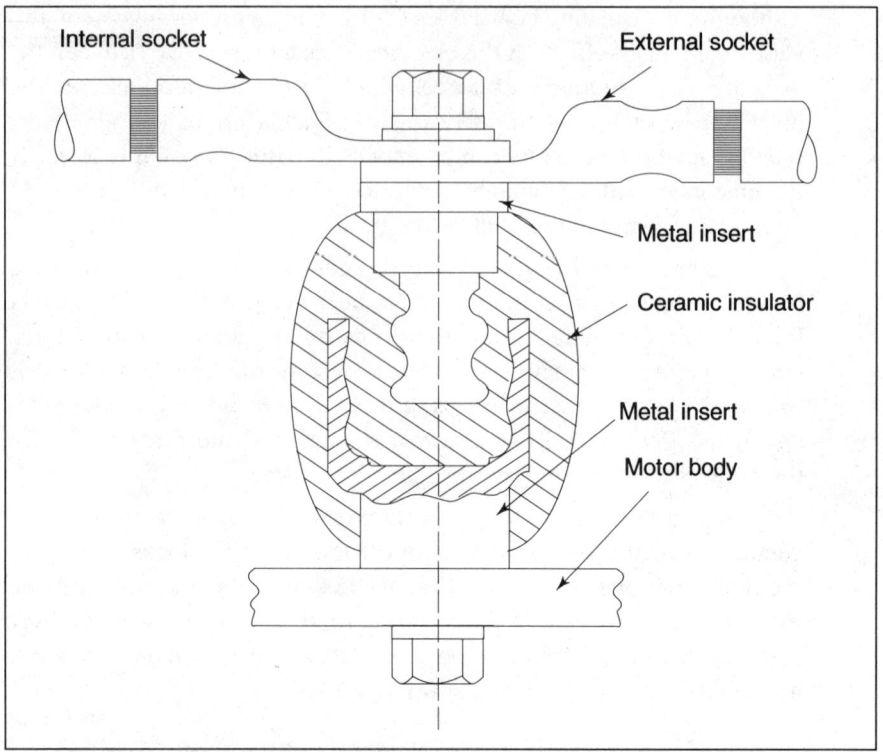

Fig. 16.4 Another improved design of bolted terminal showing terminal blocks molded into the insulation

16.4 SHRINKAGE OF INSULATION IN TRANSFORMER COILS

It may be recalled from Chapter 4 about transformer failures that one of the most common causes of interturn shorts in transformers is the shrinkage of the insulation on the conductors and the paper-based laminates which are used as spacers. The mechanisms of failure is briefly described below.

The coils are normally held under considerable pressure to prevent any relative movement between them. If and when the insulation on and between the coils shrinks, the pressure on the coils relaxes (if the counter measures described later are not taken). This enables relative movement between conductors to take place when there are any forces such as those due to vibration, electromagnetic forces caused by short circuit currents or switching in-rush currents. The relative movement between conductors is very small, but sufficient to cause progressive damage to the fragile paper insulation on the conductors. A stage is soon reached when the damaged insulation can no longer withstand even the normal voltage between turns. A short circuit between turns ensues; there is a very heavy short circuit current, gas formation and perhaps even an explosion.

The preventive measures to be taken have already been discussed in Chapter 4. They are briefly summarized here as follows:
- provision of coil pressure screws and springs,
- use of pre-compressed spacer materials,
- curing of coil stacks during manufacture to remove shrinkage,
- periodical re-tightening of the coil pressure screws.

The constructional details, the failure modes and the failure mechanisms in reactor coils are exactly the same as those in transformer coils. The preventive measures are also very similar.

16.5 SLACKENING OF INSULATED BOLTS

In many applications, bolts and nuts are insulated from the components being clamped by the use of insulating tubes and washers. Some examples of this are:
- core bolts in transformers,
- bolts used to clamp heavy busbars together,
- bolts used to clamp insulated fish-plates,
- bolts used to fix insulated bearings.

Figure 16.5 shows the general arrangement of insulated bolts and nuts. It would be obvious that any shrinkage of the insulated washers

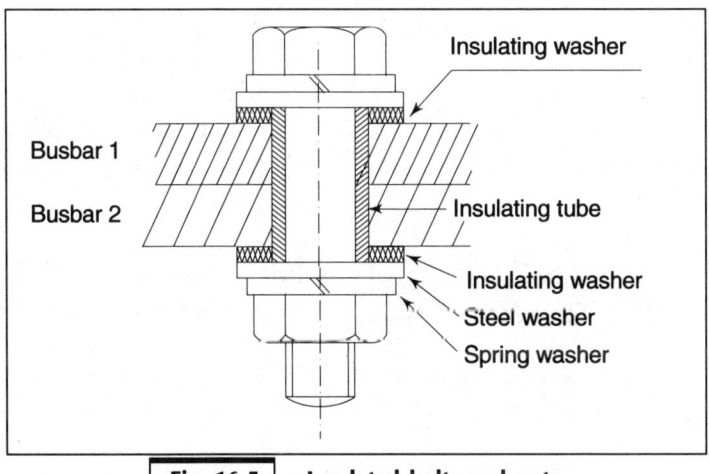

Fig. 16.5 Insulated bolts and nuts

will reduce the clamping force. The remedy is to use pre-compressed washers and to re-tighten the nuts periodically.

16.6 FIELD COIL FAILURES DUE TO SHRINKAGE OF INSULATION

In many designs of traction motors, the field coils are insulated with many layers of insulating tape bonded with varnish. Some kind of insulating sheets are also used as interturn insulation. Sometimes, insulating sheets are also placed between the coils and the yoke. All these materials are prone to shrink in service due partly to evaporation of residual solvents in the varnish bond, and partly to closing up of air gaps. As these layers of insulation are usually applied manually and the final moulding pressures are not very high, the shrinkage can be in the range of 5 to 15 percent of the aggregate initial thickness of insulation.

If the design of the pole assembly does not include adequate springs, as is the case in some designs of motors, the result of shrinkage is that the coils become loose in their setting. If the motor is subject to severe vibration, the microscopic movement of the coils in their settings is often sufficient to damage the insulation progressively until an electrical short circuit develops at normal operating voltages.

The remedy to prevent such failures is to provide springs of sufficient stiffness to ensure that the coils are held firmly against the pole piece despite the vibration. The spring deflection in the assembly is much larger than the potential shrinkage. Therefore, even when the insulation shrinks, there is only a slight reduction in the assembly force. As a corollary to this, it also follows that a defect in the spring would result in a failure of the same type as due to shrinkage of the insulation. Another design which avoids this problem altogether is one in which the coils are molded around the pole pieces with thermosetting resins. Of course this system has its own problems if there are any design or process deficiencies.

16.7 DO'S AND DON'TS FOR PREVENTING FAILURES DUE TO SHRINKAGE OF INSULATING MATERIALS

- Use only precompressed insulating materials wherever they are subject to compressive assembly forces.

- Where it is feasible to do so, introduce curing cycles of alternate heating/pressing/cooling to stabilise the assemblies during their initial manufacture.
- Introduce periodical re-tightening of the fasteners, which are likely to become loose due to shrinkage of insulating materials.
- Provide springs such as disc springs or Belleville washers in the force circuits of fasteners which are likely to become loose due to shrinkage of insulating materials.
- Modify designs of terminals similar to Fig. 16.1 as early as possible, to a type similar to one of those shown in Fig. 16.2 to 16.4.

16.8 CONCLUSION

The presence of non-metallic inserts in the force circuits of bolted electrical contacts is like a clear invitation for contact failures to occur. There have been instances of major imported equipment having this defect. In each of these cases, there were many fires and many more failures until the designs were modified on the lines indicated in Section 16.3. There were no further cases of fires or failures after these modifications were carried out.

It is suggested that every supervisor or engineer in charge of the maintenance of electrical equipment should visually examine all the terminal boards on the equipment in his charge and confirm whether there are no built-in defects in the design of these terminals. This exercise will not take more than a few hours, but it may well help to prevent failures and fires. If any terminal boards with design defects are detected, thought should be given to their modification. It is not necessary to follow the designs discussed in Sections 16.2 to 16.4. There are several other ways of achieving the same result. A method which involves the least disturbance to service can be selected. In any case, until such time as the permanent solution is implemented, the screws or nuts should be checked and re-tightened periodically. Disc springs of suitable design should be added in the force circuit.

Chapter 17

Thermal Degradation of Insulating Materials

In this chapter, we will discuss:

- The degradation of insulating properties due to the effects of operating temperature.

- Classification of insulating materials based on their ability to withstand temperature.

- Failures of insulating materials used in electrical connectors and terminal boards as a result of thermal degradation of the insulating material.

- A practical method for evaluating the condition of old cable insulation.

- Measures that can be taken to minimise insulation failures due to local overheating and thermal degradation of insulation.

17.1 INTRODUCTION

Although it is the cable insulation which burns when there are electrical fires, defective insulation is rarely the cause of fires. Failures of cable insulation are relatively rare. In most cases, the overheating or arcing that starts the fire is due to a defect in some other component or system. Cable insulation may fail in some cases, but it is very likely that in such cases, the insulation was damaged by some external condition such as mechanical, chemical or thermal damage. Under normal conditions of service, premature degradation of the cable insulation due to intrinsic defects in the quality of the insulation itself, is very rare.

Thermal Degradation of Insulating Materials

The subject of thermal degradation of electrical insulation is of great importance in the design of electrical equipment, but it is mainly important in the context of the design of the windings of the machines. Heat has the effect of making insulating materials mechanically weak and brittle. The damage is progressive and the period after which electrical failure occurs depends also on the conditions of service such as vibration and humidity. The classification of insulating materials according to their capacity of withstanding temperature is based on certain standardised test methods, the details of which are important for designers.

As far as electrical connectors are concerned, this process is relatively less important than the degradation processes discussed earlier, i.e. fatigue, creep of metals and shrinkage of non-metals. Failures of electrical connectors which are primarily due to thermal degradation of electrical insulation are very rare. Nevertheless, it needs to be discussed here mainly because efforts are often made to upgrade the insulation when it fails apparently due to thermal degradation. Actually, the observed degradation may be due to local overheating, and it may be necessary only to locate and eliminate the causes of overheating.

17.2 CHARACTERISTICS OF THERMAL DEGRADATION

It is a well-known fact that with the passage of time, most electrical insulating materials degrade or deteriorate due to internal chemical changes. It is also generally known that each type of insulating material has its own temperature class or temperature index.

The rate of degradation of any insulating material depends on its operating temperature in relation to its temperature index. If the operating temperature is equal to the index temperature, the life of the insulation is expected to be about 20,000 hours. If the operating temperature is higher than the index temperature, the life expectancy is reduced, and conversely, if the operating temperature is lower than the index temperature the life expectancy is increased. It is generally accepted that the life expectancy is doubled (or halved) for every 8° C reduction (or increase) in operating temperature with reference to the index temperature of the insulating material.

The relationship between the operating temperature and the estimated life of insulation is shown in Fig. 17.1. The very sharp or steep reduction in life as a result of excessive temperature can be seen clearly from this graph.

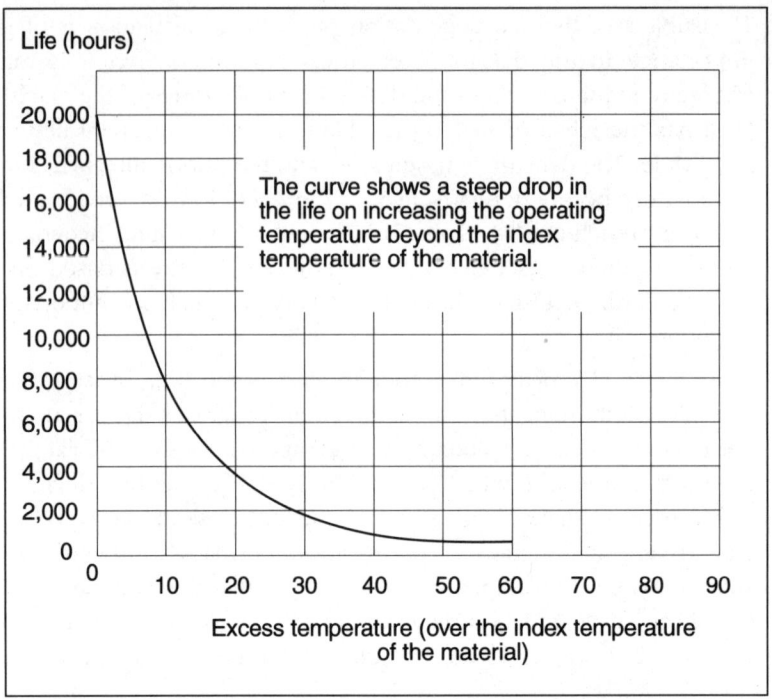

Fig. 17.1 Effect of excessive temperature on the life of insulation

The index temperature of the majority of the insulating materials in use these days is more than 100°C. (It may be noted, however, that the maximum conductor temperature for PVC insulated cables is only 70°C.) For reasons discussed in Chapter 5, the temperature of the terminals should never rise beyond 90°C. This limitation is based on the oxidation rates of copper. Thus the normal operating temperatures of terminals and sockets have necessarily to be at least 40°C lower than the temperature index of insulating materials in general use. This is the basic reason for the fact that failures of electrical connections due to thermal degradation of cable insulating materials by themselves, rarely occur close to the terminals. Such failures are more likely to occur somewhere else in the cable run, where due to proximity of other cables, or steel plates, or heat generating devices, the cable may get overheated beyond its safe limits.

17.3 INSULATION FAILURES IN ELECTRICAL CONNECTORS

Although general thermal degradation of the insulation is rarely the root cause of failures of electrical connectors and terminals, local thermal degradation may well be the effect of a defect in some other component or system, and insulation failure may be the proximate cause of failure. If there is overheating of the bolted or crimped joint, the insulating material in direct contact will get overheated, and this may be followed by electrical failure of the insulation and a resultant short circuit.

Another possible cause of thermal degradation of cable insulation and consequent failure, is the proximity effect. If a cable carrying heavy alternating currents is laid very close to a steel plate or passed through an orifice in a thick steel plate, eddy currents are produced in the steel plate and this may cause local overheating of the plate. The radiated heat from such a plate is often enough to cause failure of cable insulation. The insulation may actually ignite and the fire may spread. Whether there would be a short circuit followed by operation of automatic electrical detection devices, is then a matter of chance.

A short circuit as a result of insulation failure is often a blessing in disguise, as it may trigger a protective device and cause the power to be switched off even before a fire starts. In some cases, the attention of the operator may be drawn to the defect as a result of the tripping. Fires which may have started can be put out in their nascent stage, thereby preventing severe damages.

The relevant industry standard may be referred to for details of the tests that can be carried out to determine the suitability of the insulation and the acceptability of new cables. The most important test is the breakdown voltage test on a bent sample. This single test is the quickest way to determine the condition of old cables in service. If it is necessary to determine this condition, samples may be taken out from the installation and the test may be carried out. If the breakdown voltage is less than 50 percent of the minimum BDV specified in the standard for new cables, the cables need replacement.

17.4 DO'S AND DON'TS FOR PREVENTING THERMAL DEGRADATION OF INSULATION

- Select cable ratings after taking into account local ambient conditions and proximity of other heat generating devices.

- During service, ensure that heat dissipation from cables is not affected by accumulation of foreign materials around them.
- Consider the possibility of overheating of terminations and terminals when investigating insulation failures due to thermal degradation.

17.5 CONCLUSION

Failures of electrical connectors due to failure of electrical insulation are rare. However, failure of electrical insulation due to local overheating is often one of the effects of electrical connector failures due to other root causes.

Whenever there is any failure of insulation on or in the vicinity of electrical connectors, the possibility of overheating due to high contact resistance in the connectors, or due to proximity effect of steel plates must be examined fully before any thought is given to the question of upgrading the thermal class of the insulating material.

Chapter 18

Electrical Tracking

In this chapter, we will discuss:

- Electrical tracking, i.e. breakdown or short circuit over the surface of insulating materials, as distinct from the more common mode of electrical breakdown through the thickness of the material.

- Electrical tracking failures on terminal boards and cable ends.

- Factors which initiate and accelerate electrical tracking failures, viz. environmental conditions such as moisture, dust and chemicals deposited on the surfaces.

- Differences in the susceptibilities of different insulating materials to the tracking mode of insulation failures.

- Test methods for evaluating the resistance to tracking of insulating materials.

- Steps to be taken to minimize insulation failures through tracking.

18.1 INTRODUCTION

Most insulation failures take place through the body of the insulating material. When the voltage between conductors on either side is sufficiently high, electric current crosses the insulation through one or two 'punctures'. The insulating material gets carbonized in the puncture. If the fault current is not immediately interrupted through the operation of protective devices, the insulation around the puncture will get progressively damaged by the intense heat of the arc which is formed at the point of failure.

There is another mode of insulation failure which takes place on the surface of insulating material between conductors with a high voltage difference. In this case, the short-circuit current does not pass through the mass of the insulating material, but passes over the surface of the insulating material at a voltage which is considerably less than the voltage of flashover through air between the same terminals. Before actual failure takes place, there is often visible damage on the surface. Irregular, eroded and carbonized tracks between the two terminals are often produced. After the short circuit, the surface of the insulating material may be charred and furrowed. This type of failure is known as insulation failure through tracking. Figure 18.1 shows the difference between (a)

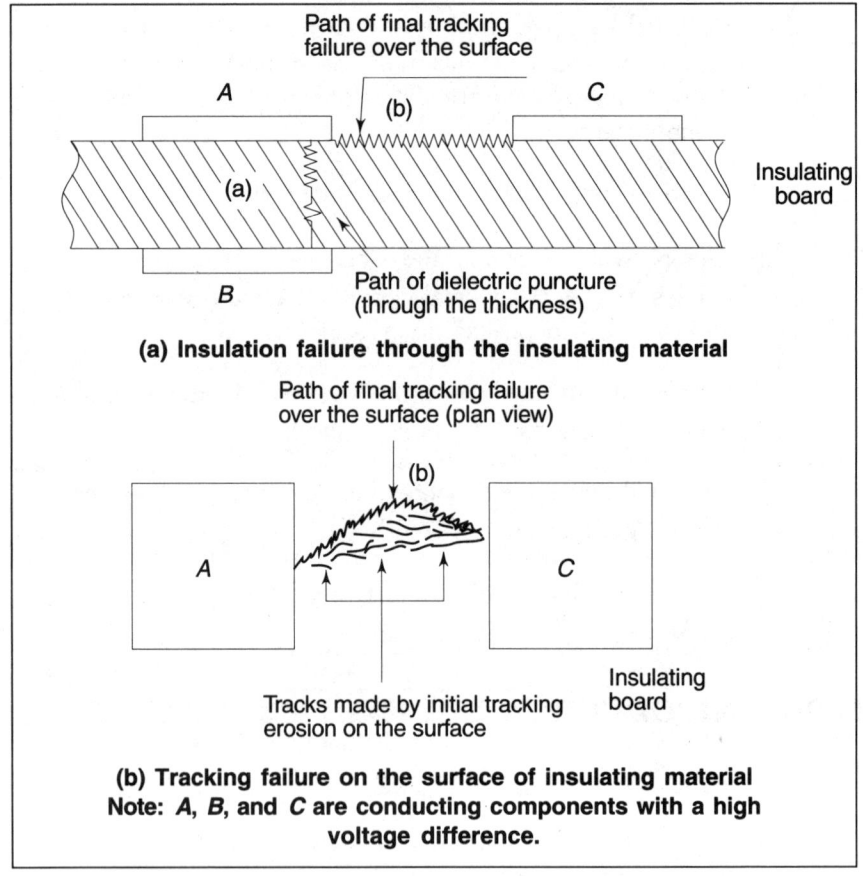

Fig. 18.1 Difference between insulation failure through material and on surface

insulation failures through the insulating material and (b) tracking failures over the surface of insulating material.

18.2 TRACKING FAILURES ON TERMINAL BOARDS

Tracking failures usually occur on terminal boards and sometimes also on cable ends.

The incidence of punctures in insulating materials depends mainly on the properties of the material; but tracking failures depend not only on the properties of the insulating material but also on environmental factors in the air such as moisture, dust, chemical salts, and oily vapors. Some insulating materials, which are specially formulated for anti-tracking properties are likely to withstand high voltages without tracking, for a longer time.

In Fig. 18.1, if the voltage between adjacent terminals A and C is excessive in relation to the shortest distance measured on the surface between these two terminals, tracking failures may take place.

18.3 TESTING OF MATERIALS FOR TRACKING PROPERTIES

It is normally not necessary to carry out tests on materials for their tracking properties when the operating voltages are less than 500 V. Most of the materials in common use have adequate resistance to tracking. However, if there are cases of failures which are obviously due to tracking and there are no signs of environmental pollution, it may be necessary to examine the possibility of the materials being deficient in their anti-tracking properties.

The testing methods and apparatus required for determining the tracking properties of materials are described in IEC 112*. The testing apparatus consists essentially of platinum electrodes of specified shapes which are made to rest on the surface to be tested with a gap of 4 mm between them. A device to apply a voltage between 100 and 600 V is connected to the electrodes. The applied voltage is increased in steps of 25 V. A solution of specified concentration of ammonium chloride and certain other additives is allowed to fall drop by drop at about one drop per 30 seconds. After every 50 drops, the voltage is increased by 25 V. This process is continued until a short

---

* IEC112 ... Standard 112 Issued by International Electrotechnical Commission.

circuit is produced. The maximum voltage which withstands 50 drops without failure is known as the *comparative tracking index* of the material. There is, of course, more to this test than the outline given here. The IEC standard should be referred to if it becomes necessary to determine the tracking index of a material.

Failures through tracking are more likely to be due to inadequate clearance and deposits of pollutants on the surface.

18.4 PREVENTION OF TRACKING FAILURES

Tracking failures can be prevented mainly by the provision of adequate distance on the surface between the terminals, proportionate to the voltage differential between them. Design details of this type are beyond the scope of this book. It may, however, be mentioned that guidance with regard to recommended distances between terminals can be obtained from standards/specifications.

A smooth and hard surface with water repellent properties between the terminals helps to minimise tracking failures. Certain anti-tracking insulating varnishes help to improve such properties. If the surface is vertical, it is less likely to collect dust and hence it will be less likely to be subject to tracking failures.

The other factors which can lead to tracking failures are environmental. Some of the more important ones which could increase the likelihood of tracking failures are as follows:

- If condensation takes place on the surface between the terminals due to sudden changes in the humidity and temperature of the surrounding atmosphere.
- If there are saline dusts or sprays in the atmosphere and they get deposited on the surface between the terminals.
- If the surface between the terminals is rough and/or dusty, it is more likely to collect condensates which could lead to such failures.

When one is confronted with failures of insulation between the terminals of a terminal board, the following possibilities should be examined:

- Did the failure start with overheating of one of the terminals due to a bad contact?
- Did the failure start with tracking between the terminals on the surface of the insulating board?
- Did the failure start with a puncture through the insulating plate between terminals, conductors or grounded structural members?

It may not be easy to decide between the three alternative failure modes, since the appearance of the failed terminal board may be the same in these three cases, unless the failure has been nipped in the bud and the early stages are visible. In case of doubt, inspection of similar equipment which have not failed may sometimes give a clue to the real failure mode. Some tests may have to be made to simulate the working conditions.

Tracking failures may take place on cable ends if a clamp or cleat is fixed too close to the end of the cable where the insulation has been stripped and the conductor is exposed. The solution is obvious. In case of outdoor cables, it is desirable to seal the ends of the stripped insulation to prevent entry of water between the layers of insulation.

18.5 DO's AND DON'Ts FOR PREVENTING FAILURES DUE TO TRACKING

- Look for signs of tracking on similar locations in other equipment before concluding tracking as the cause of a failure.
- Look for inadequate clearance and environmental pollutants before considering defective material as the possible cause of tracking.
- Ensure that surfaces liable to tracking failures are kept dry and free from pollutants by regular cleaning during maintenance schedules.
- When investigating insulation failures on terminal boards, establish whether the failure is indeed due to tracking by careful examination of the failed terminal board and, if necessary, of similar terminal boards still working in similar service. If it is so established, examine the following possibilities:
  - If the distance between the terminals is too small with reference to standard practice or in comparison with other failure-free equipment, the possibility of increasing the tracking distance by redesign of the terminal board or provision of barriers may be examined.
  - If adverse environmental factors are causing such failures, the possibility of providing counter measures such as airtight enclosures, air filters, dehumidifying chemical absorbents, vertical surfaces, and heating elements may be considered. The choice of the measures to be taken will obviously depend on the local conditions; and until such

measures are installed, periodical cleaning of the surfaces prone to tracking is the only practical method available.
- If the tracking index of the material of the terminal board is too low when measured in accordance with standard tracking index measurement methods as specified in IEC 112, the possibility of changing the material or providing anti-tracking varnish coats may be considered

18.6 CONCLUSION

Electrical short circuits or insulation breakdowns usually occur through the thickness of the insulating material. Occasionally, however, breakdowns over the surface of the insulating material are known to occur. When investigating insulation failures, it is necessary to determine which of these two failure mechanisms were initially at work, because the remedial measures would be different. This is important because the final appearance of the failed component may be the same for both these modes of failure.

Electrical tracking failures are usually due to environmental conditions, but occasionally, the cause could be either a defect in the insulating material or inadequate distance between the conductors having a high difference in voltage.

With the provision of adequate distances between electrodes, use of proper materials and periodical cleaning of vulnerable areas, it is possible to achieve zero failure performance even when the environmental conditions are very bad.

Chapter 19

Investigation of Failures and Fires

In this chapter, we will discuss:

- Systematic methods for the investigation of electrical failures and fires.
- Advantages of systematic investigation.
- Fixing priorities for undertaking investigations when many type-defects are present.
- Step-by-step approach to investigation of failures and fires.
- Additional points to be considered when investigating electrical fires.
- Drawing up, implementing and monitoring of action plans for preventing failures and fires.

19.1 INTRODUCTION

Damaged electrical installations often become the scapegoats for the cause of fires, as far as the media and sometimes even the higher management levels are concerned. While defects in electrical installations and equipment sometimes do lead to electrical fires, the serious investigator must carry out systematic investigation and not jump to conclusions at the very outset of his investigations.

Systematic investigation according to a well thought out plan helps in two ways. Firstly, the time taken is much less, and secondly, the probability of an incorrect conclusion is greatly reduced. The second reason is more important because modifications made on the basis of incorrect conclusions can be ineffective if not counter-productive, and consequently, this could be an expensive exercise.

Fires can occur due to a variety of causes. Any fire, irrespective of its cause, usually involves electrical installations, which once affected, may develop short circuits and release large amounts of energy at high temperatures, which may cause the fire to flare up further. Such cases can be said to be fires of electrical origin only if it is established that, (a) the fire would not have assumed serious proportions in the absence of the electrical energy input, and (b) the protection system against short circuits was ineffective or inoperative.

The investigation of electrical fires and failures is a responsibility which should be undertaken by the maintenance supervisors and engineers. If electrical installations are involved, it is in the interest of the electrical maintenance organisation to start an investigation immediately. Even if it is eventually determined that there was no defect in the electrical installation or equipment, there is no harm done if the equipment and system is thoroughly checked out. Maintenance engineers are in the best position to carry out investigations because they would be the first to deal with the failed equipment. Moreover, they are also the direct beneficiaries of successful preventive action determined by the investigation. Failures tend to occur repeatedly when they are due to design or manufacturing deficiencies. This can lead to adverse effects on production or service. Normalcy can be restored only when the deficiencies are identified and suitable corrective action is taken.

The best method of arresting failures is by preventing the onset of the degradation processes through the implementation of judicious changes in the design of the equipment. To be able to do so, it is first necessary to decide which of the various degradation processes are actually at work and then to examine how they can be either prevented from starting or slowed down sufficiently to prevent failures from taking place in service. Once these questions are answered, the remedial measures usually become obvious at once.

An introduction to six degradation processes which are responsible for the majority of failures of electrical equipment has already been given in the preceding chapters and for each degradation process, some practical examples have also been described in sufficient detail to enable the reader to grasp the fundamentals and to apply them effectively to investigate and to solve his own problems.

19.2 FIXING PRIORITIES FOR INVESTIGATION

In any large organisation with many types and numbers of electrical equipment in service, there may be a large number of type-failures* and type-defects which need to be investigated. It is not desirable to undertake the investigation of all such defects and failures simultaneously. If the available technical resources are spread too thinly over many problems, the results are likely to be very poor. It is better to concentrate attention on a few type-defects per engineer or senior supervisor.

Normally, it should be possible to find effective solutions to the problems taken up for study within a month or two. In the initial stages, when the engineer or supervisor is undertaking an investigation for the first time, he may be asked to take up only one problem at a time. As he gains experience and completes a few successful investigations, he can take up two or three problems simultaneously; but three problems at a time should be the limit.

It is therefore desirable to fix priorities for the investigation of failures. This can be based on the following considerations:
- Type-failures which have more serious consequences should be given priority. Thus, defects and failures which have the potential of causing fires should be given the highest priority.
- Type-failures which have very serious effects on production or services should get the next priority.
- When fires and failures have been covered and fully investigated, defects may be taken up for study. (Chapter 2 may be referred to for recalling the difference between defects and failures.)
- Within the types classified above, type failures or type-defects which have a higher repetition rate should get higher priority.

Since in any case all the failures have to be eliminated totally, it is not difficult to decide the priorities. Often it is possible for engineers in charge of maintenance to draw up a list of priorities based on their experience and a brief reference to failure statistics.

As soon as the type-failures to be investigated are assigned to different engineers and supervisors, it is possible for them to focus their attentions to the assigned projects. The remaining paragraphs in this chapter are addressed to those who take up the systematic

---
* A type-failure is one in which there are three or more failures with the same failure mode, and the same failure mechanism.

investigation of type failures and fires. We will discuss the steps assuming a single project. It is possible to follow exactly the same process for one or two more projects simultaneously.

19.3 PROCESS OF INVESTIGATION

When one has gained enough experience in the investigation of failures, it is often possible to pass over some of the steps discussed below; but for the inexperienced investigator, it is desirable to go according to the suggested method, which is infallible.

It is desirable but not essential to follow the sequence of steps given below. Depending upon the availability of equipment, documentation, concerned engineers and staff, the sequence may be changed as convenient; but it is desirable to take all the suggested steps before coming to any final conclusion about the cause and the remedial measures.

All equipment operate according to immutable laws of nature and all failures or fires also occur strictly according to these laws. There are no exceptions and there is never any stage in the failure mechanism which does not follow some natural law. It is only necessary to discover the true failure mechanism and once that is done, it is always possible to devise suitable changes in the design, manufacture, operation or maintenance as a result of which the failures will stop completely. Those investigating failures of electrical equipment can do so with total confidence that if they follow logical steps of investigation, they are sure to reach the correct solutions.

19.4 FAILURE INVESTIGATION: STEP-BY-STEP APPROACH

- It is important to follow a systematic approach for failure investigation. The following steps can serve as guidelines for a step-by-step approach: The first step in failure investigation is to open a file or folder for each type defect or type failure being taken up for study. All data and documentation collected from time to time should be placed in the file folder. This step is very important. It helps the investigator to put things in perspective, and puts the investigation along the right lines. Since the investigator would be able to spend very little time every day on the project, it helps to maintain continuity and saves time. Even if at the end of a

Investigation of Failures and Fires 257

reasonable effort, no solution is found, it is possible to let the matter rest and start off again with fresh vigor after a break.
- Collect and tabulate full details of each failure of the type selected for study during the preceding year. The details to be tabulated should include names and makes of components which failed first; date, time, place, and load when the failure occurred; dates of installation, previous maintenance schedules and repairs; name of operator; and every other possible detail that can be collected with reference to the incident and the component/equipment involved. Often, but not always, such a tabulation will throw up common factors that may point the way to a correct solution as regards the mechanism and cause of failure.
- Examine the failed components and failed equipment. Do so even if there is considerable damage. Try to reason out which component failed first. Try to differentiate between cause and effect. In general, all damages which are clearly the effects of the failure can be ignored, and attention concentrated on the components which failed first. If there is any doubt about one or more components, consider them all as the likely source until more data is collected and more thinking done. Keep notes about the observations. If possible, take photographs, specially if the defective item is to be reused after repair.
- Make a list of the probable causes of failures of the components considered to be the likely source of the problems. Examine similar components in other equipment in service. It is possible that you may notice incipient signs of failure in some of these components.
- Study the following documents very carefully with special reference to the components involved in the failure:
  - the operating and maintenance manual
  - the manufacturing drawings
  - failure reports and photographs, if any, of previous incidents
- Study previous failure investigation reports on the same subject if any such investigations were made; but do not accept the conclusions and recommendations immediately. Keep an open mind until all the data are collected and then accept only those conclusions which fit in with all the available data.
- Discuss the failures with others in the field who may have knowledge or experience of the subject under consideration, but again, do not come to any conclusions until you have collected all the data and done your own thinking.

- Compare the experience of other installations where similar, or preferably, identical equipment are in service. If you find some place where there are no failures of the type under consideration, try to find the differences in environmental, operating, manufacturing or maintenance conditions.
- Consider also the observations made during scheduled maintenance of these equipment. If necessary, make your own observations. Keep the list of probable causes of failures in mind.
- Study the operating conditions during actual service. Look for unusual features or conditions which do not conform to the manufacturer's operating instructions.
- By this time, you would have come to tentative conclusions regarding the possible mechanisms of failure and the contributory causes. Verify these hypotheses on equipment in service. If necessary, make the required measurements. If the hypotheses are confirmed, then decide on the modifications necessary to avert the failures.
- Every defect is a potential failure and every failure starts as a defect. Hence, it is desirable to monitor the incidence and analyse the statistics of defects in the components under investigation. Defects are usually detected during routine maintenance and inspection. Defect statistics should, therefore, be maintained and classified for detailed study of related defects when investigating the failures of any particular type of defect. Those types of defects which are more likely to cause failures and fires should be given priority for detailed examination. The following questions may be asked and answered for each type of defect which is related to the type-failure under investigation.
  - At which stage was the defect introduced? Design, manufacture or maintenance?
  - Which is the earliest stage at which the defect can be detected?
  - Is any improvement possible or necessary in the method of detecting such a defect?
  - Is it necessary to improve the design or process of manufacture to minimise such defects?
  - Is the required know-how and know-why being imparted to the concerned staff in a systematic training schedule?
  - Does the relevant documentation (drawings, specifications, process sheets, training material, quality assurance schedules) cover the important aspects relating to the prevention of such defects?

19.5 REVIEW OF FAILURE STATISTICS

It must be ensured while making any comparisons, that the failure statistics are considered for the same period. Further, if there are any differences in the conditions of service such as loading, environment, or supply voltage, the failure rates would not be strictly comparable. There may be differences in the failure modes. Despite such limitations, it is desirable to maintain statistics of failure rates of different types and makes of electrical equipment and components. When wide variations are noticed between different groups, it is necessary to determine the reasons for the same. The reasons could be in specification, design, manufacture, operation, maintenance, environment, and service conditions. Such investigations may help to determine the ways in which failure rates can be reduced in any particular group showing a high failure rate.

If the hours worked per year of different equipment are widely different, failure rates can be calculated on a per hour or per thousand-hour basis instead of per year basis. But then, the hourly utilization statistics have to be maintained for all equipment.

19.6 INVESTIGATION OF ELECTRICAL FIRES

Since electrical fires always start as failures of electrical equipment, the methods for investigation suggested in Section 19.4 apply fully for the investigation of electrical fires too. However, there are a few additional points relevant to electrical fires.

Whenever there is an electrical fire, it is necessary to locate the starting point of the fire. This is most important. Generally, everything around the starting point of the fire is destroyed (i.e., burnt or incinerated). If the fire is not controlled quickly, vast areas may be affected. It is often felt that it is impossible to determine the starting point of a fire. Some suggestions which may help in locating the starting point are:

- Evidence from those who first detected the fire may help to localize the starting point of the fire to some extent.
- If the switchboard, panelboard, or motor control center (MCC) has not been damaged, the condition of the overcurrent protective devices (i.e., fuses and/or circuit breakers) should be checked. If high interrupting capacity fuses have operated, it is desirable to send them to testing laboratories for analysis. It is possible to determine, by X-ray examination, whether the fuses

operated on overload or short circuit. Similarly, if there are overcurrent, ground fault, and other protective devices, they should be checked to see whether any of them have operated. If some of them have operated, it may help to localize the source of the fault. The electrical and physical locations of the circuits protected by the devices which have operated should be checked against the appropriate plans, risers, or other wiring diagrams.

- The entire copper or aluminium wiring and electrical equipment should be checked inch by inch to locate those portions which are molten and congealed, which are covered by globules of metal, or which show other signs of arcing. Compared to other signs, melting of the conductors is more likely to be associated with the starting point of the fire. In general, the heat developed by the fire is not adequate to melt copper. Hence, any evidence of molten copper can be associated with an electrical arc. In the case of aluminium conductors, aluminium oxide powder may be detected because aluminium can combine with oxygen at arc temperatures. All such residues should be collected, labelled and analyzed.
- If any molten wires or components of equipment are found, their position on the electrical schematic diagram should be marked, as this might help to explain the sequence of events and the possible source of fire.
- If the fuse boards or protective device boards are intact but have not operated, all the devices should be checked to see if any of them are defective, i.e., which fail to operate under fault conditions.
- Operators' log books should be perused to see if there are any abnormal indications which might be relevant to the occurrence of a fire.

The observations mentioned above should be made as early as possible after the fire has been extinguished, before any of the evidence is disturbed. Damages which are clearly the effects of the fire should be recorded separately. As far as the cause is concerned, in the case of electrical fires, it is likely to be found in a small section of the damaged zone. Molten metal conductors (or aluminium oxide powder, where aluminium conductors are used) are very likely to be at the starting point of the fire. Sometimes however, fires can start in the insulating material due to the overheating of contacts and conductors, and short circuits and molten conductors may be merely the effects.

Investigation of Failures and Fires

Efforts should be made to find a hypothesis which fits into all the observed facts.

Electrical fires are more likely to be due to the failures of electrical connectors and terminal boards because there are no protective systems to detect them. Failures due to short circuits are less likely because protective systems are usually provided, and if a fire still occurs, it would be due to a defect in the protective system.

The measures or precautions to be taken to prevent failures of electrical connectors and terminals are very simple, inexpensive and practical. If such measures are not being taken in some organizations, it is only due to lack of awareness of the significance of the facts.

Finally, it must be emphasised that it is not sufficient if the senior officers and supervisors alone are knowledgeable in these matters. It is the staff who actually do the work, who should be trained about the correct and incorrect practices. Mareover, it is also not sufficient if the majority of such staff are well trained—it is necessary to ensure that hundred per cent of the staff are fully trained to do the work correctly. Electrical connectors are all fitted manually and it is usually not practicable to test or inspect each and every one of the thousands of locations in each installation. It is necessary that the staff who do the work do it right every time.

It follows that a hundred percent of the staff who are assigned this work should be well trained for that purpose through systematic, practical, modular training courses. Therefore, during investigations of failures and fires, one very important point to be checked out is whether the staff concerned are fully aware of the correct operating and maintenance instructions.

19.7 ACTION PLAN

The end product of any investigation has to be an action plan—a plan which will completely stop the incidence of failures. the action plan may require implementation in any one or more of the following areas:
- specification
- design
- manufacture
- installation
- operation
- maintenance

If action needs to be taken in either operation or maintenance, the action plan can be implemented more easily. If, however, action is needed in specification, design, manufacture or installation, the plan would be more difficult to implement. In any case, it would not be relevant for the equipment already installed. In such cases, it is necessary to have an additional separate plan for the equipment already installed and in service.

It is obvious that investigation and even the determination of the preventive measures will have no effect in the field until the action plan is implemented for all the equipment in service. This may involve purchase of new materials, and modifications on existing equipment. The last stage is the monitoring of the equipment to confirm that the failures have really stopped occurring.

19.8 CONCLUSION

The methods outlined above would enable any individual or organization to get to grips with the problem of failures and fires in electrical equipment. A beginning can be made at any time. It is not necessary to wait for a devastating fire to occur before embarking on such a study. In fact, such an occasion would not arise if pre-emptive steps are taken—the right time for which is when everything seems to be apparently under control.

A great deal of effort is involved in the investigation of failures, determination of the preventive measures and the implementation of the action plan; but it is all worth doing because the rewards are much greater.

Appendix 1

⚡ FIRES INVOLVING ELECTRICAL EQUIPMENT BUT NOT CLASSIFIED AS ELECTRICAL FIRES

As stated at the outset, this book is mainly about fires and failures caused by defects introduced in electrical equipment due to deficiencies in design details, manufacture and maintenance of electrical equipment. In other words, it is about incidents for which responsibility rests with the electrical engineers in charge. It is meant mainly for the electrical engineers, supervisors and technicians at the front lines.

This book deals with aspects which seem to have received little attention in published literature, but which are responsible for many cases of electrical fires. In this Appendix, causes of fires involving electrical equipment, but which cannot be considered to be fires originating from defects in electrical equipments or installations are enumerated. For instance, if a careless user places a radiant heater facing a curtain made of flammable or combustible material, the resulting fire could be said to have started from an electrical appliance, but it was obviously due not to any defect in the appliance but due to carelessness or misuse. Similar accidents could result in the following cases:

- An electrical pressing iron is left in the ON condition flat on an ironing board.
- A fan type room heater is kept ON, but some item of apparel made of cotton or synthetic material is dropped on the heater, blocking the air flow.
- A 2000-W hot-plate is connected to a 15-A wall receptical. This could result in repeated blowing of the branch fuse, which may have been 'solved' by using a higher-rated fuse.
- A 1500-W immersion heater is switched ON in a plastic bucket full of water and the user forgets to switch it off on time.

- An electric kettle used for making tea in an office, is filled with water, switched ON and then forgotten. An accident is inevitable if the kettle is kept with office files.
- A welder does electric arc welding in the well of an elevator shaft, where oil drips from the elevator guide rails, and rags, scrap paper etc. have accumulated over the years into a mass of flammable material.
- An electric hairdryer is switched ON, kept on the bed, blocking the air entry grill and forgotten.

Each of the eight cases described above could result in a fire accident. The number of ways in which electrical appliances can be misused is very large. A great deal of energy can be released from an electrical appliance or from an overloaded wire. While the majority of users are careful in installing, using and handling electrical appliances, mistakes of the type mentioned above are also common, generally due to ignorance, and sometimes due to plain carelessness.

Some of the appliances using heaters are provided with built-in thermostats or thermal cut outs which cut off the power supply to the heater if it gets overheated due to any reason. However, this is no guarantee of safety because such devices can become defective. In the ultimate analysis, there is no defence against misuse of electrical appliances. Training of electricians and educating the users of electricity are therefore the most important steps that need to be taken to minimise fires and accidents.

Appendix 2

⚡ TECHNICAL MEASURES FOR PREVENTING ELECTRICAL FIRES AND MINIMIZING THE DAMAGES

The emphasis in this book is on the prevention of seed-defects and defects which could lead to electrical fires. There are many possible technical features and devices, apart from the mandatory protective devices, which could help to prevent or detect electrical fires, or which could minimise the damages resulting from electrical fires. This aspect has not been dealt with in this book. New innovations are constantly coming into the market and some of these are indeed quite effective. There are others which are unreliable and which only give a false sense of security. Further, the kind of seed defects and defects discussed in this book can and often do occur even in these devices. Therefore, all these should be evaluated and used with care. Some of these devices are enumerated below.

Fire Resistant Cables In general, the insulation and protective sheaths on electrical conductors are made of combustible substances like PVC, natural or synthetic rubbers, and other organic or synthetic materials. When an electrical fire starts due to overheating of conductors/joints or sparking/arcing, the first thing to ignite is usually the insulation on the electrical wires. When the insulation burns, sulphurous or acidic fumes are often produced. Some of these are very toxic. So far, no one has developed an insulating material which matches the electrical, mechanical and cost-effective properties of these insulating materials, but which is totally fire resistant. Efforts in that direction are, continuing, and some materials with improved fire resistance properties are being advertised from time to time. However, they are usually more expensive than the conventional insulating materials.

Fire Protection Coatings of various types are being marketed. It is claimed that they help to reduce the rate of flame propagation along cable insulation.

Fire Seal Systems are used to seal openings in walls through which electrical cables are passed. They help to contain the fire to the room where it originates and to prevent it from spreading rapidly.

It does, however, seem much cheaper and more sensible to take the few steps necessary at the cable terminations and supports to prevent the starting of the fire, rather than to make an attempt to slow down the propagation of the fire. The measures suggested in the preceding chapters are not only less expensive, but also more effective than the use of expensive materials which do not prevent fires but merely promise to reduce the damage.

Earth Leakage or Current Balance Relays are more sensitive than conventional overcurrent relays. They are capable of detecting and aborting fires in their nascent stages.

Thermostats are often provided in electrical appliances which incorporate heaters, in order to prevent fire, despite misuse by the users. These thermostats detect a rise in temperature and cut off the power supply before any damage can occur to the equipment, or before a fire can start. Such safety devices are not legally mandatory, and appliances without such devices are commonly sold in the market.

Fireproof Chokes for fluorescent tubes are available, but there are others in the market which may catch fire either due to an internal fault or due to a defect in the tube or starter.

Capacitors are intrinsically prone to fail sooner or later due to the very high electrical stresses in their dielectrics. Some makes of capacitors are, in fact, provided with internal fuses which operate and disconnect from the supply, any capacitor which develops a short circuit within its dielectric, and some types of capacitors have self-healing dielectrics. Some makes of capacitors explode when they fail and this may start a fire. There seems to be no legal requirement to prevent marketing of devices which can catch fire when they fail in service due to any reason. It is necessary for the careful user to insist on devices which are fail-safe. It would perhaps be unrealistic to ask for a device which would never fail, but it is certainly possible and hence desirable to specify that the device must not catch fire under any circumstances. The stipulation should then be verified by type tests in which random samples are made to fail by the application of a high voltage.

Dry Type Transformers have become available. They have the advantage of being free from large quantities of combustible oil. Similarly, transformers using *Synthetic Non-Combustible Fluids* instead of oil are also available. While these may be considered on the basis of their individual merits for future installations, the precautions described in Chapter 4 must be taken for dealing with the large number of conventional transformers already in service, or those which may be installed in future on account of their lower cost.

Busducts or Busbar Trunk distribution systems eliminate the use of cables, and to that extent, the fire hazard from the use of combustible cable insulation is reduced. These systems have other advantages too. Their use may be considered on overall merits and economics. It must be reiterated here, though, that very inexpensive measures at the terminations can ensure total freedom from fires when cables are used, and the same measures have to be taken even when busducts are used in order to prevent failures.

Selected Readings and Terms

There are hundreds of books on the various subjects introduced in this book. Many of them are highly theoretical and they are unlikely to generate any interest amongst the supervisors and junior engineers who are at the front lines, so to speak, of the campaign against failures of electrical equipments.

A very small list of references is given below. These are mostly Standards and Handbooks which are usually available in any good technical library. Interested readers who would like to study more about the topics mentioned below may make a beginning with these very practical and useful publications.

Crimping British Standard 6360 gives practical details of endurance tests on crimped sockets. These are a must when selecting the design, dimensions and manufacturer of crimping sockets and crimping tools. These are for copper wires.

Indian Standard 8337 gives similar information for aluminium conductors.

Reverse Bend Test This is a very simple but effective test for checking the mechanical strength of copper wires and strips. Indian Standard 1716 gives full details of not only the test, but also the device which is needed to carry out the test. Brittleness in copper wires due to hydrogen embrittlement can be detected by this test.

Fatigue *The Metals Handbook* published by the American Society for Metals gives very useful information regarding failures of metal components due to various failure mechanisms in Volume 11 with the title 'Failure Analysis and Prevention'. Hundreds of illustrated case studies are given here. Although failures of crimped sockets or wire strands are not mentioned, the information is very relevant.

Selected Readings and Terms

Creep Volume 8 of the *Metals Handbook* gives details of the behaviour of metals at elevated temperatures (i.e., Creep).

Electrical Tracking It is necessary to be very careful about every little detail of the testing procedure. Reference may be made to IEC 112 for full details regarding the apparatus to be used and the procedure for testing.

Reliability Engineering Those who are interested in a study of Reliability Engineering may refer to the following books which are especially suitable for beginners.

Practical Reliability Engineering by Patrick O'Connor, published by John Wiley.

A Practical Approach to Reliability Engineering by R.H. Caplan, Published by Business Books Ltd.

Transformers *The J & P Handbook on Transformers* is a very practical and useful reference book which deals with every important issue relating to the use of power transformers. The chapter on failure of transformers is essential reading for those investigating transformer failures.

Electric Fuses by A Wright and P.G. Newbery, published by Peter Peregrinus Ltd. This book would be useful to those interested in the selection and testing of electric fuses.

Bolted Joint Design by William Eccles in Engineering Design of November 1984 gives practical and theoretical guidance on the selection and use of threaded fasteners for different types of applications.

Selection and use of Engineering Materials by F.A.A. Crane and J.A. Charles, published by Butterworths is a very useful book for engineers to understand the full potential and limitations of commonly used engineering materials.

Indian Standards are now available for almost all the materials, equipment and hardware used in the electrical industry. There are also many Codes of Practice which give guidance on maintenance practices. Reference should be made to the latest volume of the *Handbook of Standards*, which is published periodically by the Bureau of Indian Standards, Bahadur Shah Zafar Marg, New Delhi. Very often, a number of standards are available for similar products. The most appropriate standard should be selected after checking the scope of the available standards. Sometimes, Indian Standards may not be available for certain items. In such cases, reference may be made to the *Handbook of British Standards*.

It is often necessary to refer to a number of specialised standards for getting all the relevant details relating to specifications, test methods, and test equipment. All the required cross references will be found in the main standard for the equipment or material under consideration.

A brief list of Indian Standards which would be useful in the context of the subjects dealt with in this book is given below:

IS 302	Safety requirements for household appliances
IS 398	Aluminium conductors for overhead lines
IS 434	Rubber insulated cables
IS 7391	Copper conductors
IS 694	PVC cables
IS 867	Testing of moulded materials
IS 1271	Thermal classification and evaluation of insulating materials
IS 9224	Low voltage fuses
IS 8187	D-type fuses
IS 2329	Bend test on metallic tubes
IS 2824	Tracking index of solid insulating materials
IS 1403	Reverse bend test
IS 1576	Solid press-board for electrical purposes
IS 5074	Fatigue test (Rotating bar type)
IS 5619	Fatigue testing of metals
IS 5780	Intrinsically safe electrical equipment
IS 8504	Thermal endurance properties of electrical insulating materials
IS 8828	Miniature circuit breakers for less than 1000 V
IS 9926	Fuse wire in rewirable fuses
IS 1255	Code of practice (COP) for power cables
IS 732	COP for electrical wiring
IS 1866	COP for mineral insulating oils
IS 10028	COP for installation and maintenance of transformers

Index

Note: boldface numbers indicate illustrations

Action plan to prevent failures, 261
Alloys, 7
Alternating and fluctuation stresses in, 206–207, **207**
Aluminum, 7
Approach of this book, 5–8, **5**
Arcing
    Connectors and terminals, 92
    Crimped sockets, 140
    Electrical fires and, 35, 36, 38, 41, 42, 43, **43**, 52, 55

Bearings, failures in, 25, 26–27
Bend testing
    Brazed, soldered, welded joints, 189, **189**
    Crimped sockets, 141–142, **141**
    Reverse bend test, 268
Bolted connections
    Failure due to metal creep, 217–220, **218**, **219**
    Shrinkage in non-metallic materials, 239–240
Brazed, soldered, welded joints, 186–197
    Applications and uses for, 187, 96–197
    Bend test for, 189, **189**
    Brazed joints, 191–194, **193**, **194**, 196
    Butt welded joints, 187, 188–190, 195

    Connectors and terminals using, 98, 186–197
    Correct methods for, 195
    Soldered joints, 194–195, 196
    Temperature limits in, 187
    TIG welded joints, 187, 190–191, **191**, 196
    Tinning or silver-plating of surfaces for, 187
Busbars, 57, 267
Busducts, 267
Butt welded joints, 187, 188–190, 195

Cable (*see* **wiring and cables**)
Capacitors, 266
    Damage to, 26
Carelessness as cause of fires, 37
Ceramic insulators, damage to, 26
Ceramics, 7
Chemical damage to insulation, 161, 167
Chokes, fireproof, 266
Circuit breakers, 21, 48, 51, 54, 59
    Electrical fires vs., 35, **39**
    Jammed, 22
Cleanliness of materials and contact resistance, 226, 228
Clearance for bearings, 25
Coatings, fire protection coatings, 266
Codes, standards, 48–49, 269–270
Coil stack in transformers, 71, **72**

Combustibles (*see* **flammable materials**)
Commonality of systems, failures, 6–7
Comparative tracking index, 250
Compressive stress/strain, 31–33, **32**
   Metal fatigue and, 201, **201**, **202**
Conductivity of copper in crimped sockets, 141
Conductors
   Damage to, 22, 26
   Fractured or broken, as cause of fire, 8, 17, 38–41, 45, 55–57
   Transformers and, 71
Connectors and terminals (*see also* **crimped sockets; plug & socket connectors**), 9, 57
   Applications for, 86–89, **87**
   Arcing caused by, 92
   Bolted connections, **88**, 88
   Brazed, soldered, welded joint failure in, 98, 186–197
   Causes of failures in, 94–95
   Clamp type, 45
   Contact force in, 95
   Criticality of failures in, 93
   Damage to, 22
   Defects as cause of failures in, 57, 85–103
   Design defects in, 91, 98–99
   Detection of failures in, 91
   Effects of failure in, 92
   Failure of, 10–11, 15, 45
   Fracture of crimped socket for, 97, 104–121
   Installation precautions for, 94
   Insulation failure in, 97, 159–169, 159
   Investigation of failures in, 93
   Mechanisms of failure in, 96–98
   Modes of failure in, 96–98
   Overheating and, 97, 99–102, 104–121
   Plug & socket type, **89**, 122–138
   Prevention of failure in, 93–94, 102
   Screwed or nut/bolt type, 95–96
   Seed-defects in, 91
   Short circuits in, 92
   Spring-held types, 96
   Temperature limits for, 91–92, 100–102
   Terminal board failure and, 97
   Transformers failures and, 64, 68–79
   Typical types of, 86–89, **88**, **89**
   Wire strand fracture in, 97
Contact force
   Connectors and terminals, 95
   Contact resistance and, 226–228, **226**
   Plug & socket connectors and, 127–128
   Shrinkage in non-metallic materials and, 235
   Terminal boards and, 180–182
Contact resistance, 12, 224–232
   Cleanliness of materials and, 226, 228
   Contact force and, 226–228, **226**
   Contact theory and, 229–231, **230**
   Cross section of contact point, 227, **227**
   Factors influencing, 225–229
   Fasteners and, 229
   Importance of, 225
   Normal of materials and, 226
   Points of contact in common connections, 224–225
   Prevention of failure in, 231–232
   Resistivity of materials and, 226
   Surface area and, 228–229
   Yield point of materials and, 226, 227
Contact theory, 229–231, **230**
Control panel, 9
Copper, 7
Core bolt insulation failure, transformers, 78–79
Cracking due to metal fatigue, 199
Creep in metals (*see* **metal creep**)
Creep temperature limit (CTL), 30–31, 215–216, **216**
Crimped sockets, 57, 104–121
   Application technique for, 109–111, **110**, **111**

Index

Calculating crimping reduction for, 114, 115
Connector and terminals using, 104–121
Cross section of, 114, **114**
Cut wires and, 154–156, **155**
Cyclic process of failure in, 112–115, **113**
Endurance testing of, 116, **116**, 117
Fractures of, 139–149
—arcing in, 140
—bend testing of, 141–142, **141**
—bends and breaking in, 142–144, **143**
—causes of, 139–140
—conductivity of copper and, 141
—cracking in, 143–144
—defects and, 139, 140–142
—endurance limit of metal in, 143
—investigation of, 147–148
—metal fatigue in, 140
—prevention of, 148–149
—stress and, 139, 142–144, **143**
—stress concentration factor in, 143
—tensile strength in, 142
—thermal stress and, 139, 146–147
—vibration and, 139, 144–146, **145, 146, 147**
—yield point in, 142
High contact resistance in, 113–114
Inspection of, 117
Lug-type, 106–109, **107**
Modes and mechanisms of failure in, 111–115
Overheating and burning of, 104–121
Overheating of, due to terminal board failures, 119
Prevention of failure in, 115–116, 120–121
Resistance in, 116
Socket system of terminating cables using, 106, **107**
Temperature testing of, 116
Tin-solder type, 106–109, **107, 108**

Wire working out of, 119–120, **120**
Crimping standards, 268
Current balance relays, 266
Current ratings, 59

Defects and deficiencies (*see also* seed defects), 4–5, 8, 10, 11, 14, 15, 17, 22, 24–25, 258
Connector and terminal failure caused by, 89–92, 91, 98–99
Crimped socket failure caused by, 139, 140–142
Electrical fires caused by, 41
Plug & socket connector failure caused by, 137–138
Protective system failure caused by, 49–50
Seed-defects (*see* seed defects)
Terminal board failures caused by, 170
Transformer failure caused by, 67, 70
Wire strand failure caused by, 158
Deformation, plastic, 26
Deterioration or degradation of materials, 12–14, 16–17, 41–44
Insulation and, 47–48
Dielectric strength
Insulation, 34
Megger test of, 34
Dissolved gas analysis (DGA), transformers, 69
Dry type transformers, 267

Earth leakage, 266
Elastic limit of materials, 33
Elastic stress, 26
Elasticity (*see also* metal creep), 30
Metal creep, 213
Young's Modulus of elasticity and, 213
Electrical contact resistance (*see* contact resistance)
Electrical damage to insulation, 161, 167–168

Electrical fires, 21, 35–60
  Arcing and, 35, 36, 38, 41–43, **43**, 52, 55
  Carelessness and, 37
  Causes of, 35–38, 40–45, 56
  Circuit breakers vs., 35, **39**, 48
  Conductor fractures and, 38–41, 45, 55–57
  Consequences of, 40
  Defects and, 41
  Degradation of materials and, 44
  Deterioration of materials and, 41
  Fires involving electrical equipment but not classified as, 263–264
  Flammable materials and, 38, 52
  Fuses vs., **39**, 48
  Incorrect installation and, 40
  Insulation failure and, 35, 38–41, 43, **43**, 44, 45–48
  Investigation of, 40, 259–261
  Misuse of electrical appliances/ equipment and, 38
  Monitoring equipment to prevent, 58
  Motors and, 36
  National Electric Code (NEC) vs., 48–49
  Overheating and, 36, 38, 41, **42**, 43, **43**, 55
  Overloading and, 38–41, 41, **42**, 43, **43**
  Parallel clamp failure and, 36, **37**
  Pressure contact failure and, 38–41, 43, **43**, 44–45, 55–57
  Prevention of, 57–58, 265–267
  Protection systems vs., 39, 48, 49–55
  Protective system failure and, 36, 37
  Reactors failure and, 37
  Severity of, 38
  Short circuits and, **36**, 37, 41, 43, **43**
  Sparking and, 38, 41–42, 52
  Technical measures for prevention of, 265–267
  Temporary wiring and, 37
  Transformers and, 36
  Transmission lines and, 36
  Zero failure performance vs., 38
Endurance limit of materials, 33
  Crimped sockets and, 143
  Metal fatigue and, 208
Expansion, thermal, 26
Explosive materials (see flammable materials)

Failures, 3–4, **4**, 7
  Action plan to prevent, 261
  Connectors and, 15
  Definition of, 21–22
  Degradation processes and, 12–13, 14, 16–17
  Design defects and, 11, 14, 15, 17
  Electrical contact resistance and, 12
  Incorrect installation and, 11, 14, 15
  Insulation degradation and, 12, 17
  Investigation of, 13–14, 15, 18, 25–27
  Maintenance vs., 12, 17, 18
  Management role in prevention of, 14–16, 18
  Manufacturing defects and, 11, 14, 15
  Materials and, 15
  Mechanisms of, 25–27, 67
  Metal creep and, 12
  Metal fatigue and, 12, 198–211
  Modes of, 25–27, 66–68
  Overloading and, 17
  Prevention of, 13, 15
  Protective systems and, 54–55
  Quality control vs., 15–16
  Rates of, calculations for, 27–28
  Reasons for, 11–13, 14, 15
  Seed-defects and, 17, 18, 19, **20**, 23–25
  Short circuits and, 17
  Shrinkage and, 12
  Terminals and, 15
  Thermal degradation and, 12
  Tracking, electrical, and, 12
  Zero failure performance and, 15, 17, 18, 23, 25
Fasteners
  Contact resistance in, 229

Index

Loosened, in terminal boards, 172–175, **173**, **174**
Loosened or damaged, 22
Fatigue of materials (see metal fatigue)
Field coils, shrinkage of insulation in, 240
Fire prevention measures, 57–58
Fire protection coatings, 266
Fire resistant cables, 265
Fire seal systems, 266
Fireproof chokes, 266
Flammable materials (see also lubricants), 38, 52, 59
Fluctuating stresses, 206–207, **207**
Force circuits, terminal boards, 177
Fractures of crimped sockets (see crimped sockets, fractures of)
Fuses, 39, 48, 52–53, 59, 269

Gas tungsten arc welding (GTAW) (see TIG welded joints)

Heat-shrink sleeves to reduce flexing in wires, 152–154, **153**, **154**
Heating effect of current, 26
High-potential (hipot) tests for insulation, 47

Index temperatures and thermal degradation, 244
Indian Standards, 59, 269–270
Installation, incorrect, 11, 14, 15
  Codes vs. 48–49
  Electrical fires and, 40
Insulation
  Causes of failure in, 17, 46
  Chemical damage to, 161, 167
  Connector cables, failure of insulation in, 159–169
  Connectors and terminals and, 97
  Damage to, 22, 26
  Defect or deficiency of design in, 46–47
  Deterioration or degradation of, 12, 47–48
  Dielectric strength of, 34
  Electrical damage to, 161, 167–168
  Electrical fires caused by failure of, 8, 35, 38–41, 43, **43**, 44–48
  Fuses and circuit breakers in failure of, 48
  High-potential (hipot) tests on, 47
  Installation defects and, 46
  Integrity of, 49
  Mechanical damage to, 160, 161–165, **161**, **162**, **163**, **164**, **165**
  Megger test for, 34
  NEC rules for, 49
  NRTL testing and certification in, 46
  Overheating and, 165–167
  Plug & socket connectors and, 135
  Prevention of failure in, 50–51, 168–169
  Protective systems and, 49–55
  Resistance of, 34
  Shrinkage of, around transformer coil, 238–239
  Shrinkage of, field coils, 240
  Thermal damage to, 160, 165–167
  Tracking in, 247–249, **248**
  Transformers and, 67, 71, 75–77
  Types of, 46
  Vibration and mechanical damage in, 161–165, **161**, **162**, **163**, **164**, **165**
Interrupt current ratings, 49
Interturn shorts, transformers and, 70–77
Investigating failures and fires, 13–14, 15, 18, 25–27, 40, 60, 253–262
  Action plan to prevent failures, 261
  Defects and, 258
  Degradation processes in, 254
  Electrical fires, 259–261
  Fires involving electrical equipment but not classified as electric fires, 263–264

Priorities in, 255–256
Process of, 256
Statistics of failure in, 259
Step-by-step approach to, 256–258
Iron and steel, 7

Laminated layers and tracking in transformers, 70, **70**
Laminates, 7
Load, 5, 59
Lubricants, 25, 27
   Synthetic non-combustible fluids vs. lubricants, 267
Lug-type sockets, 106–109, **107**

Maintenance as preventive of failures and fires, 3–4, 6, 12, 13, 17, 18
Management role in prevention of failures, 14–16, 18
Manufacturing defects (see defects and deficiencies)
Material properties, 7, 15
   Compressive stress/strain, 31–33, **32**
   Creep temperature limit, 30–31
   Elastic limit, 33
   Elasticity, 30
   Endurance limit, 33
   Melting point and creep temperature limit (CTL), 215–216
   Metal creep, 12, 30–31, 45, 212–223, 269
   Metal fatigue, 12, 29, 140, 198–211, 268
   Shrinkage in non-metallic materials, 233–241
   Stress and strain, 31–33, **32**
   Yield point in, 29
Mechanical damage to insulation, 160, 161–165, **161**, **162**, **163**, **164**, **165**
Mechanisms of failure, 25–27
   Connectors and terminals, 96–98
   Crimped sockets, 111–115
   Shrinkage in non-metallic materials, 233–234
   Transformers and, 67

Megger test of insulation resistance/dielectric strength, 34
Melting point and creep temperature limit (CTL), 215–216
Metal creep, 12, 30–31, 45, 212–223, 269
   Bolted connection failure due to, 217–220, **218**, **219**
   Creep temperature limit (CTL), 215–216, **216**
   Elasticity and, 213
   Electrical contacts and, 214–216
   Failures of connectors due to, 217–220
   Melting point vs., 215
   Parallel clamp failure due to, 221–222, **222**
   Prevention of failure caused by, 220–221, **221**, 222–223
   Rate of, 213–214, **214**, **215**, 216
   Screwed connections failure due to, 219–220, **220**
   Soldered joint failure due to, 217
   Temperatures vs., 212
   Tensile stress and, 212–213
   Terminal board failures and, 181, 183–184
   Time and, 213–214, **214**, **215**, 216
   Young's Modulus of elasticity and, 213
Metal fatigue, 12, 198–211, 268
   Alternating and fluctuation stresses in, 206–207, **207**
   Causes of, 208–209
   Compressive stress and, 201, **201**, **202**
   Cracks in, 199
   Crimped sockets, 140
   Endurance limit of metal and, 208
   Failures caused by, 198–199
   Fractures caused by, 204–206
   Identifying fractures caused by, 209–210
   Nature of, 207–208
   Prevention of, 210–211
   S–N curve showing, 204–206, **205**

Index

Stress concentration factor (SCF) and, 201–203, **203**, 206
Tensile stress and, 199–201, **200**, **202**
Wire strand failure caused by, 151, 156, **157**, 156
Miniature circuit breakers (MCB), 59
Misuse of electrical appliances/ equipment, 8, 38, 263–264
Modes of failure, 25–27
    Connectors and terminals, 96–98
    Crimped sockets, 111–115
    Plug & socket connectors, 132–133
    Transformers, 66–68
Moisture and tracking, 249
Monitoring equipment, fire prevention, 58, 59
Motors, electrical fires and, 36
Multicore coupler using plug & socket connectors, 123–125, **123**, **125**, 131–132, 136

National Electric Code (NEC), 48–49, 51
National Fire Protection Association, 48
Nationally Recognized Testing Laboratory, 46
Neglect of maintenance, 3–4
Normal of materials and contact resistance, 226

Open circuits, plug & socket connectors and, 133, 135–136
Overcurrent protection, 51, 54
Overheating
    Connectors and terminals and, 97, 99–102, 104–121
    Electrical fires and, 36, 38, 41, **42**, 43, **43**, 55
    Insulation damage and, 165–167
    Plug & socket connectors, 129–130, **130**, 132, 133–135
    Transformers and, 71, 80–82
Overloaded circuits, 2, 17
    Electrical fires and, 38–41, 41, **42**, 43, **43**

Transformers and, 63, 64–66
Overvoltage in transformers, 80–82
Oxidation, 26

Parallel clamp failure
    Electrical fires caused by, 36, **37**
    Failure due to metal creep, 221–222, **222**
Plastic deformation, 26
Plastics, 7
Plug & socket connectors, 122–138
    Applications and uses for, 122–123
    Causes of failure in, 128–131, **130**, **131**
    Clamping to prevent pull-out in, 136, **137**
    Contact force in, 127–128
    Defects in, 137–138
    Electronic devices using, 136
    Industrial multicore couplers and, 131–132
    Inspection of, 138
    Installation techniques for, 137
    Insulation failure in, 135
    Loosening of threaded terminal block from pin in, 130–131, **131**
    Maintenance of, 123
    Modes of failure in, 132–133
    Multicore coupler using, 123–125, **123**, **125**, 131–132, 136
    Open circuits in, 133, 135–136
    Overheating in, 129–130, **130**, 132, 133–135
    Prevention of failure in, 138
    Pull-out force test for, 126, **126**, 134–135
    Seed-defects in, 137–138
    Short circuits in, 133
    Specifications for, 125–126
    Spring arrangements used in, 127–128, **127**
    Three-pin types, in domestic and general use, 128–131, **129**
    Voltage and current ratings in, 123, 125–126

Pressure contact failure
　Defects causing, 17
　Electrical fires and, 38–41, 43, **43**, 44–45, 55–57
Preventive maintenance, 3–4, 7, 8, 13

Protective systems (*see also* **circuit breakers; fuses**), 24, 59
　Calibration of, 55
　Circuit breakers as, 51, 54
　Defects in, 49–50
　Electrical fires and, 36–39, 48, 49–55
　Failures of, 21
　Fuses in, 52–53
　Maintenance of, 55
　Miniature circuit breakers in, 54
　National Electric Code (NEC) and, 51
　Overcurrent protection in, 51, 54
　Preventing failures in, 54–55
　Short circuits and, 51–52
　Simple system, block diagram of, 51–52, **52**
Pull-out force test for plug & socket connectors, 126, **126**, 134–135

Quality control and failure prevention, 15–16

Rate of failure, calculations for, 27–28
Reactor failure, electrical fires and, 37
Reasons for electrical equipment failure, 11–13, 14–15
Relays
　Earth leakage or current balance type, 266
　Jammed, 22
Reliability engineering, 6, 269
Repairs, 3–4, 7
Resistance (*see also* **contact resistance**)
　Insulation, 34
　Megger test of, 34
Resistivity of materials and contact resistance, 226

Resistors, damage to, 26
Reverse bend test, 268

Screwed or nut/bolt connectors, 95–96
Seed-defects (see also defects and deficiencies), 17–19, **20**, 23–25
　Connectors and terminals, 91
　Plug & socket connector failure caused by, 137–138
　Terminal board failure caused by, 170
　Transformer failure caused by, 67, 70
　Wire strand failure caused by, 158
Seizure of bearings, 27
Shafts, 26, 27
Short circuits, 51
　As cause of fires, 2, 8, 17, **36**, 37, 41, 43, **43**
　Connectors and terminals failure, 92
　Plug & socket connector and, 133
　Protective systems and, 51–52
　Transformers and, 66, 70–77
Shrinkage and failure, 12, 233–241
　Contact force and, 235
　Field coil insulation and, 240
　Insulated bolt slackening due to, 239–240
　Insulation in transformer coils and, 238–239
　Mechanisms of failure and, 233–234
　Prevention of failures due to, 240–241
　Terminal design and, 234–237, **234**, **236**, **237**, **238**
　Terminal board failure caused by, 175–178, **177**
S–N curve for metal fatigue, 204–206, **205**
Soldered joints (see also brazed, soldered, welded joints)
　Metal creep and failure of, 217
Sparking, electrical fires and, 38, 41–42, 52

Index

Spring washers used in terminal boards, 182–183, **182**
Spring-held connectors, 96
Statistics of failures, 7–8
Strain (see stress and strain)
Strength, 5
Stress and strain, 31–33, 32
   Alternating and fluctuation stresses in, 206–207, **207**
   Crimped sockets, 139, 142–144, **143**
   Elastic stress in, 26
   Metal fatigue and, 199–201, **200**, **202**
   S–N curve for metal fatigue, 204–206, **205**
   Stress concentration factor (SCF) and, 143, 201–203, **203**, 206
   Wire strand failure, 156, **157**
Stress concentration factor (SCF), 143, 201–203, **203**, 206
Switchboard panels, 21
Synthetic non-combustible fluids vs. lubricants, 267

Tapchanger in transformers, 69–70
Temperature, temperature limits
   Brazed, soldered, welded joints and, 187
   Connectors and terminals and, 91–92, 100–102
   Creep temperature limit (CTL), 215–216, **216**
   Metal creep vs., 212
   Terminal boards and, 180, 181
   Thermal degradation and, 243–244, **244**
Temporary wiring, electrical fires and, 37
Tensile strength/tensile stress
   Crimped sockets, 142
   Metal creep and, 212–213
   Metal fatigue and, 199–201, **200**, **202**
   Wire strand failure and, 156, **157**
Terminal boards, 170–185
   Causes of failure in, 171
   Connector and terminal failures in, 97
   Contact force in, 180–182
   Defective design example in, 175, **175**
   Defects in, 170
   Designing for reliability in, 178–180, **178**, **180**
   Force circuits in, 177
   Inadequately sized terminals in, 172
   Loose fasteners in, 172–175, **173**, **174**
   Metal creep in, 181, 183–184
   Prevention of failure in, 184–185
   Seed-defects in, 170
   Shrinkage and, 175–178, **177**
   Spring washers used in, 182–183, **182**
   Temperature and shrinkage in, 176–178, **177**
   Temperatures causing failure in, 180, 181
   Thermal stress and, 181
   Tracking as cause of failure in, 183, 249
   Typical design of, 171, **171**
Terminals (see connectors and terminals)
Terminology, 19–34
Thermal damage to insulation, 75–77, 160, 165–167
Thermal degradation, 12, 242–246
   Characteristics of, 243–244
   Failures caused by, 245
   Index temperatures and, 244
   Insulation and, 242–246
   Operating temperatures vs., 243–244, **244**
   Prevention of, 245–246
   Rate of, 243–244
Thermal expansion, 26
Thermal stress
   Crimped sockets and, 139, 146–147
   Terminal boards and, 181
Thermostats, 266
Three-pin plug and sockets, in domestic and general use, 128–131, **129**

TIG welded joints, 187, 190–191, **191**, 196
Tin-solder sockets, 106–109, **107**, **108**
Tinning or silver-plating of surfaces for brazed, soldered, welded joints, 187
Tracking, 12, 247–252, 269
    Comparative tracking index and, 250
    Insulation failure and, 247–249, **248**
    Moisture and, 249
    Prevention of, 250–252
    Terminal board failure and, 183, 249
    Testing materials for, 249–250
Transformers, 61–84, 269
    Applications for, 62
    Capacities/ratings of, 62
    Coil stack in, 71, **72**
    Conductors in, 71
    Connector failures within, 64, 68–79
    Core bolt insulation failure, 78–79
    Defects in, 67, 70
    Dissolved gas analysis (DGA) in, 69
    Dry type, 267
    Electrical fires and, 36
    Failure mechanisms in, 67
    Failure modes in, 66–68
    Failure rates in, 62
    Failures of, 61–84
    Insulation failure between winding and tank, 77–78
    Insulation failure in, 67, 71, 75–77
    Interturn shorts and, 70–77
    Investigating failures in, 82–83
    Laminated layer and tracking in, 70, **70**
    Laminations, insulation failure between, 79–80
    Life-span vs. temperature in, 65–66
    Overheating in, 71, 80–82
    Overloading in, 63, 64–66
    Overvoltage in, 80–82
    Prevention of failure in, 83–84
    Procuring new, 82–83
    Relative movement of turns and interturn shorts in, 73–75, **74**
    Seed-defects in, 67, 70
    Short circuits in, 66, 70–77
    Shrinkage of coil insulation in, 238–239
    Sizes of, 61–62
    Synthetic non-combustible fluids vs. lubricants, 267
    Tapchanger in, 69–70
    Terminal failures within, 64, 68–79
    Thermal damage to insulation in, 75–77
    Typical power network using, 62, **63**
    Weather-related failures and, 63
    Windings in, 71
Transmission lines, electrical fires and, 36
Tungsten inert gas (see TIG welded joints), 190

Vibration
    Crimped socket failure caused by, 139, 144–146, **145**, **146**, **147**
    Insulation damage caused by, 161–165, **161**, **162**, **163**, **164**, **165**
    Wire strand failure caused by, 151, 152–154, **153**, **154**
Vicious cycle of failure and neglect, 3–4, **4**

Welded joints (see brazed, soldered, welded joints; TIG welded joints)
Wiring and cables
    Causes of failure in, 151–152
    Connectors and terminals for, 97
    Crimped sockets cutting in, 154–156, **155**, 154
    Defects in, 151, 158
    Excessive flexing in, 151, 152–154, **153**, **154**
    Failures of, 150–158
    Fatigue in, 151, 156, **157**
    Fire protection coatings for, 266
    Fire resistant types of, 265
    Fractures in, 57
    Fractures inside insulating sleeves of, 156, **156**, 156

Heat-shrink sleeves to reduce flexing in, 152–154, **153**, **154**
Incorrect size of, 2
NEC rules for, 49
Prevention of failures in, 157
Seed-defects in, 158
Sharp edges and cutting in, 151, 154–156, **155**
Size of, 2
Tensile stress in, 156, **157**
Vibration and, 151, 152–154, **153**, **154**

Yield point of materials, 29
    Contact resistance, 226, 227
    Crimped sockets, 142
Young's Modulus of elasticity and metal creep, 213

Zero failure performance, 5, 15, 17, 18, 23, 25, 38

About the Author

A. A. Hattangadi is an electrical reliability consultant specializing in prevention and investigation of electrical fires and failures, with a diverse international industrial clientele. An electrical engineer, he is former general manager of the Cittarianjan Locomotive Works. He also conducts training courses for the Indian Railways Institute.